LITERACY IN SCIENCE, TECHNOLOGY, AND THE LANGUAGE ARTS

LITERACY IN SCIENCE, TECHNOLOGY, AND THE LANGUAGE ARTS

An Interdisciplinary Inquiry

Mary Hamm and Dennis Adams

BERGIN & GARVEY
Westport, Connecticut • London

Library of Congress Cataloging-in-Publication Data

Hamm, Mary.
  Literacy in science, technology, and the language arts : an interdisciplinary inquiry / Mary Hamm, Dennis Adams.
    p.   cm.
  Includes bibliographical references and index.
  ISBN 0-89789-575-4 (alk. paper). — ISBN 0-89789-576-2 (pbk. : alk. paper)
    1. Science—Study and teaching. 2. Engineering—Study and teaching. 3. Language arts—Study and teaching. 4. Group work in education. 5. Literacy. I. Adams, Dennis A., 1958- .
  II. Title.
  Q181.H215   1998
  507.1—DC21   98-3300

British Library Cataloguing in Publication Data is available.

Copyright © 1998 by Mary Hamm and Dennis Adams

All rights reserved. No portion of this book may be reproduced, by any process or technique, without the express written consent of the publisher.

Library of Congress Catalog Card Number: 98-3300
ISBN: 0-89789-575-4
       0-89789-576-2 (pbk.)

First published in 1998

Bergin & Garvey, 88 Post Road West, Westport, CT 06881
An imprint of Greenwood Publishing Group, Inc.

Printed in the United States of America

The paper used in this book complies with the Permanent Paper Standard issued by the National Information Standards Organization (Z39.48-1984).

P

Copyright Acknowledgments

The authors and publisher gratefully acknowledge permission for use of the following material:

Excerpt from "January First" by Octavio Paz, translated by Elizabeth Bishop, from *The Complete Poems 1927-1979* by Elizabeth Bishop. Copyright © 1979, 1983 by Alice Helen Methfessel. Reprinted by permission of Farrar, Straus & Giroux, Inc. (Originally published in *The New Yorker*.)

"The Summer Day" by Mary Oliver from *House of Light* in *New and Selected Poems*. Used with permission of Beacon Press.

Every reasonable effort has been made to trace the owners of copyright materials in this book, but in some instances this has proven impossible. The authors and publisher will be glad to receive information leading to more complete acknowledgments in subsequent printings of the book and in the meantime extend their apologies for any omissions.

Contents

	Preface	vii
1	Interdisciplinary Inquiry with Science, Language, and Literacy	1
2	Promoting Critical and Creative Thinking	17
3	Cooperative Learning: Teamwork and Individual Accountability	33
4	Assessing Student Understanding	53
5	Scientific Literacy: Involving Students in Scientific Inquiry	67
6	The Language Arts and Literacy	95
7	Interdisciplinary Themes in Science and the Language Arts	129
8	Educational Technology: The Multiple Possibilities of Powerful Tools	157
	References	189
	Index	197

In order to keep this title in print and available to the academic community, this edition was produced using digital reprint technology in a relatively short print run. This would not have been attainable using traditional methods. Although the cover has been changed from its original appearance, the text remains the same and all materials and methods used still conform to the highest book-making standards.

Preface

At one time, literacy was squeezed into an established framework of reading and writing. The meaning of literacy has changed as new circumstances have opened up a much wider range of possibilities. New communications technology is transforming teaching and changing how we come to understand the natural world. Seeing, hearing, writing, and reading with a multimedia computer changes the way we approach a subject and the whole dynamic of communication.

What will it mean to be literate in the next century? This book will attempt to shed light on that question as it concentrates on weaving the concept of literacy into the teaching of science and the language arts. Technology, a product of science, is pushing against the linear boundaries of traditional storytelling. Moving in the direction of multiform stories and digital formats takes literacy well beyond the 3Rs. Students increasingly need to be critical and creative users of this new medium. Even the formats imposed on the computer by older media are fading into the background as human communication is becoming less bound by time, space, and form. The Internet is a good example. As it becomes faster, more visually powerful, and easier to manipulate, there will be an explosion of virtual environments. Literacy takes on a whole new meaning as these parallel universes become more like a populated landscape than a billboard-strewn highway.

The word *literacy* has become almost synonymous with the word *competence*. We like to stay a little closer to the more traditional definitions found in the new language-arts and science standards. The National Science Education Standards (National Research Council,

1995) suggest that "scientific literacy implies that a person can identify scientific issues underlying national and local decisions and express positions that are scientifically and technologically informed."

Because of its central focus on the language arts, science, and technology, this book will include many aspects of science education and educational technology under the literacy umbrella. Recognizing the role of the scientific endeavor and understanding how science interacts with society is one of the basic dimensions of scientific literacy. The National Council on Science and Technology Education identifies a scientifically literate person as one who does the following:

1. Recognizes the diversity and unity of the natural world.
2. Understands the important concepts and principles of science.
3. Is aware of the ways that science, technology, language, and other core subjects depend on each other.
4. Knows that science and technology are human endeavors and recognizes what this implies about their strengths and weaknesses.
5. Has a capacity for scientific ways of thinking.
6. Makes use of scientific knowledge and ways of thinking in personal and social interactions (American Association for the Advancement of Science, 1990).

Scientific literacy includes seeing scientific endeavors through the perspective of cultural and intellectual history and being familiar with how these ideas cut across subject boundaries. People have always been concerned with transmitting attitudes, shared values, and ways of thinking to the next generation. Today it seems more important than ever because every part of contemporary life is bombarded by science and technology. Part of scientific literacy also consists of clarifying attitudes, possessing certain scientific values, and making informed judgments. Accepting this challenge, we invite students to use the whole range of language skills to enhance and cultivate scientific patterns of thinking, logical reasoning, curiosity, an openness to new ideas, and skepticism in evaluating claims and arguments.

We recognize the fact that language-rich home and school environments have traditionally been the key to literacy. As the concept of literacy extends to meet the demands of today's complex world, there is little question about the increased importance of science, technology, media, and a wide range of language activities. It is our hope that over time teachers will become more open to the possibilities that appear at the intersection of science, language, and literacy.

When it comes to reading and the other language arts, the term literacy is now often used to cover a wide range of complex tasks required to communicate. The new standards in the language arts include things like visual literacy and communications technology.

As this book explores important new dimensions of linguistic and scientific literacy, it looks at developing literacies not well covered in school today. It is our belief that an understanding of science and the processes of science can make major contributions to the ability to learn, reason, make decisions, and solve problems. Learning about the natural world helps develop intellectual tools of inquiry that can be used with the language arts and other disciplines. Interdisciplinary activities can help develop skills in both science and language that are becoming so essential for personal fulfillment, the workplace, and citizenship. Subject-matter integration and thematic teaching are just two of the possibilities suggested for injecting broad literacy-based concepts across the curriculum.

This book is designed as a text for general-methods classes in curriculum and instruction. It can serve as a supplemental text for science education or an integrated language-arts class. It is also intended as a resource book and guide for preservice and inservice teachers. Interdisciplinary inquiry requires knowledgeable teachers who set up time for exploration, observation, experimentation, inquiry, and the full range of literacy-related activities. Teachers are viewed as the key ingredient because they are the ones who must structure the learning environment on a day-to-day basis. Nothing could be more important than assisting them with their continuous professional development.

The authors wish to thank book producer John Beck for all his help and Wei Wei Cai for her important contribution to the section on thematic units.

LITERACY IN SCIENCE, TECHNOLOGY, AND THE LANGUAGE ARTS

Chapter 1

Interdisciplinary Inquiry with Science, Language, and Literacy

> *Literacy is more than an individual's ability to read, write and speak in English. It also involves inquiry, computing and solving problems at levels of proficiency necessary to function on the job and in society, to achieve one's goals, and to develop one's knowledge and potential.*
> —National Literacy Act (1991)

This book examines the characteristics of effective literacy instruction and what it will mean to be a literate person in tomorrow's world. Curriculum elements are viewed through the wide lens of a new literacy. This chapter takes a thematic approach that focuses on scientific inquiry, communications media, experiental everyday examples, and integrated problem-solving activities. Centering learning around the deep themes that underlie content makes it hard to separate things out by subject.

Whether it is science or the language arts, students have to have experiences in the classroom that model the realities of the world outside of school. And that world does not place things neatly into categories. Planning more integrated literacy lessons requires attention to children's real-world interests and needs. Students will profit from exploring the real ways adults actually use language and information technology. They will also profit from interdisciplinary inquiry with scientific ideas related to health, populations, resources, and environments. The whole process provides the building blocks for student understanding and action as citizens.

It is our belief that children learn best when they construct knowledge from their own experiences. This constructivist approach is based on the view that knowledge cannot be gained simply by absorption through the senses, but by active thinking and doing. Learning any subject involves the continuous reshaping of the mental processes that mediate learning. As children do this they combine what they already know with new information. From the center to the new outer boundaries of literacy children must experience learning and be allowed to build meanings and relationships for themselves. Clearly, learning is something students do. It is not something that is done to them.

In opening a bigger literacy umbrella, we hope to challenge your thinking and further stimulate your interest an integrated curriculum. No matter how well we arrange and connect the present, there will always be new things happening, new information coming in, new worlds to explore, and a constantly expanding domain of life, learning, and consciousness. A literacy-intensive interdisciplinary curriculum simply reflects the fact that change really does favor the prepared mind.

CONNECTING TO THE WORLD OUTSIDE SCHOOL

A thorough understanding of language and literacy is essential for learning any subject, because it makes the world more understandable, more interesting, and more connected to daily life. All students need to comprehend the new dimensions of literacy and how they relate to their culture and their lives. As the new national standards for the language arts points out, "Goals lack meaning if students are not motivated to integrate their knowledge willingly, effectively, and joyfully into their academic work and into their lives outside the classroom" (National Council of Teachers of English [NCTE], 1996, p. 44).

As students examine reality from many angles and in different lights, they engage in social, physical, and mental activities that allow them to visualize new connections and choices. The boundaries between disciplines should not be closed borders. What it takes to be truly literate in today's world requires moving beyond the usual walls between subjects. An integrated curriculum, built around literacy skills, is a powerful way of teaching and organizing an instructional program. The interdisciplinary interweaving of skills and processes is an open invitation for connecting learning in ways that tie in with prior experience. Implementation is a dynamic process that takes time, commitment, and patience. It serves as a form of inquiry and serious professional development for the teachers involved.

Many of the scientific views held today resulted from many small discoveries over time and are a product of cultural and historical ways of thinking and viewing the world. Some have come from rather sudden paradigm shifts, as new data force scientists to see the world differently. Knowledge building is not always a smooth progression. Significant historical events, such as Galileo's perspective on the earth's place in the universe, Newton's discoveries of laws of motion, Darwin's observations of evolving life forms, and Pasteur's identification of infectious disease organisms, are milestones in the development of Western thought. Fruitful scientific thinking has always involved the free interplay of concepts and sense impressions. This is just another way of saying theory and observation. We suggest adding a third element: maintaining a sense of wonder in the face of nature.

> Language makes us human
> Literacy makes us civilized
> The arts add enlightenment and enduring resources for thoughtfulness
> Science and technology make us powerful
> And being in community with others can make us free

SUBJECT MATTER, EFFECTIVE INSTRUCTION, AND THE POWER OF GOOD STORIES

The construction of meaning around big ideas requires connecting important elements of the curriculum. The abundance of activities suggested throughout this book will build on this theme and place it under the umbrella of literacy. Pedagogical and content-area knowledge are viewed as equally important. Without a solid subject-matter base, teachers will find it difficult to teach content and provide appropriate feedback. Interdisciplinary work requires a foundation in the disciplines involved. Without a knowledge of pedagogy it is difficult to manage a class or make a subject meaningful and interesting for students. Just because a teacher knows a subject well does not mean that he or she can teach it.

Traditionally, there has been a gap between what was taught and what was really learned. Interpreting and understanding the world—and how it relates to personal experience—is different than the interpretations and facts found in traditional lessons. In the past, many school programs have produced students with increasingly negative attitudes about most of the subjects that they are studying. This is less likely to happen when lessons are broadly focused and the teacher considers the needs, interests, motivations, and experiences of the learners.

When you are working with almost any subject you are working with a way of understanding and knowing about the world. In this book we often suggest literature as a way to make all subjects come alive. By using storybooks, children can question and use thinking skills to organize their experiences and observations. Literature helps to keep science from being squeezed into narrow boundaries. Literature can show students how science relates to their personal, social, and imaginative life. Good stories also make science more palatable for children.

SHIFTING INSTRUCTIONAL REALITIES

Significant efforts and a change in thinking are needed if we are going to create a citizenry that is truly literate. Teaching subjects in isolation leads to isolated thinking and less frequent use in real-world situations. Exploring the big ideas that link important areas of knowledge requires a sound grounding in science, technology, language, and literacy. These forces are reciprocal in nature.

Teaching children and preparing them for the future requires shifting the definition of literacy to include learning about reality from many angles. As children use literacy skills to work with ideas and symbols from across the curriculum, they learn to identify their assumptions, use critical and logical thinking, and consider alternative explanations. As one question leads to another, inquiry ripples outward to form ever-larger concepts and reveal ever-more universal ideas about communication, our lives, and our world. Whether it is creatively solving problems, thinking critically, communicating effectively, working cooperatively, or making good use of technology, students need to be active and engaged with the curriculum and each other.

A paradigm shift often begins with new discoveries and experimental discrepancies that cannot be squeezed into the usual framework. As new discoveries about learning and new approaches to knowledge building come into their own, we are forced to view literacy in a different light. It is our view that implementing an interdisciplinary curriculum under an umbrella of literacy requires time, motivation, shared beliefs, a developmental process, and participants who share ownership of the new possibilities being opened up. Literacy education now requires thinking critically, working collaboratively, communicating clearly, and sharing in the excitement of comprehending the natural world.

The new standards for the language arts and the new science education standards call for a more inclusive literacy. Both go on to suggest that students must participate as knowledgeable, reflective, creative, and critical members of a variety of literacy communities. If we are to enter a more mature and more productive period in human history, it is important for all students to become literate in a manner that allows them to flourish in the 21st century. Change is not always a smooth or linear process; it is interactive, dynamic, complex, and strongly influenced by environmental factors. Putting new subject-matter standards into practice requires well-prepared teachers and school policies that support the vision contained in the standards.

Connecting inquiry-based science to work in the language arts is a structure for learning that can enhance both subjects. Such interdisciplinary inquiry is becoming part of good educational practice, especially in areas where students have limited English proficiency. Teachers can articulate the central ideas of multiple disciplines by using broadly based themes as a vehicle for gaining understanding across multiple subjects. Many teachers now use a thematic approach to explore overarching concepts or big ideas that cross discipline boundaries.

Traditionally, subjects have been taught as single disciplines at the same time every day. Sometimes this results in bits and pieces of isolated and fragmented information. Interdisciplinary inquiry can focus on specific topics like magnetism, electricity, or life cycles. Breaking down some of the barriers between subjects also allows teachers to develop units around broader themes, such as the technology revolution or the dwindling rain forest. Once they are used to working with themes, students can take part in choosing interesting topics. Ideas can come from what they are studying, community issues, the newspaper, literature, film, video, or the Internet.

Scientific literacy allows individuals to read and understand articles about science in the popular press. Students who are scientifically literate are able to identify scientific issues and make decisions that are scientifically informed. In addition, they understand scientific principles and processes well enough to discuss, pose, and evaluate scientific arguments on the basis of evidence. Becoming scientifically literate requires a wide range of communication skills (like reading) and an understanding of information technology. We view science, language, and literacy as correlational themes that can enhance real-world problem solving. Investigating important community or global issues puts students in a context where they have to communicate, reason, and collaborate. Thus, real-world problem solving provides a rich context for integrating the curriculum.

Today's citizens are increasingly called upon to make decisions regarding problems that are not limited by discipline boundaries. This means participating effectively in decisions ranging from funding electrical power plants to approving the distribution of genetically engineered life-forms. Our fast-paced technological and communications culture also requires workers who can go beyond machine calculations to critically think and collaboratively solve problems related to real-world situations. It is little wonder that interdisciplinary inquiry is part of both the science and language-arts standards. In a world filled with the products of language, science, and technology, a broadly defined literacy has become a necessity for everyone. We all need to use the skills of language, science, and information technology to make choices that arise every day.

All students should be offered the opportunities, the encouragement, and the vision to develop the language skills they need to pursue life's goals, including personal enrichment and participation as informed members of our society. More than ever, everyone must understand the basic demands of literacy and science to grasp patterns, solve problems, and deal with the ambiguity of a constantly changing world. Literacy growth begins before children enter school and continues after they graduate. As the new standards point out, the language arts is more than reading and writing. It also includes speaking, listening, graphic representation, and related technological resources. Students are now expected to be knowledgeable, reflective, creative, and critical members of a variety of literacy communities.

Science and language are closely related systems of thought that naturally correlate to each other and the natural world. Communications technology itself stems from both subjects and can provide students with concrete examples of language and scientific processes in real-world situations. Being skilled in the language arts enables students to achieve deeper understanding of scientific concepts. It also provides them with tools for quantifying and explaining scientific relationships.

INQUIRY-BASED PROBLEM SOLVING
WITH SCIENCE AND LANGUAGE

Though children need to learn how to construct knowledge for themselves, it should be recognized that they frequently have many false ideas about science and literacy applications. Whether misconceptions are natural or learned, all have to be dealt with in a way that makes room for the beauty, quirkiness, and complexity of science, language learning, and their associates. What this means for the classroom teacher is that he or she has to figure out how to make science and language (which can often be abstract and intimidating) comprehensible and relevant to the everyday world of children.

A well-designed curriculum can serve everybody (including those with special needs) and still be committed to a common core of learning in language arts and science. All our students can solve problems creatively, think critically, and work cooperatively. Becoming knowledgeable about the biological and physical universe is impossible if you do not have the capacity to accomplish a wide range of communication skills.

The process requires teachers who can help students feel actively involved, motivated, and reasonably competent. Since we are living on an incredible roller coaster of technological change, it is important that in the future the communications technology and science be more in harmony with itself, its users, and the curriculum.

We invite you to think about the following:

- How children reason with literacy expectations and how they learn from direct experience with real things.
- How an integrated approach to scientific and linguistic literacy can invigorate both.
- How to help students reason and collaborate as they investigate language, science, and the tools of technology.
- How the skills of language and the scientific processes can help in making the leap from facts that can be observed to complex realities that cannot be experienced.
- How you can make inquiry-based science and active language critique believable, meaningful, and pleasurable for you and your students.

A surprising percentage of the American population is unreceptive to all or parts of science. Many adults also fear new technology. In spite of agreeing to the importance of these subjects, much of the public has trouble overcoming their innate fears. Newspapers frequently report alternative approaches, like the woman in Chicago who tied her electrical cords in knots to reduce her utility bills (Raloff, 1996). This may not be a bad experimental procedure, but the result was predictable: It did not help much. For teachers to move students beyond socially constructed fictions and math/science phobias involves looking for the language connections that provide insight into our subjects, ourselves, and the world around us.

LANGUAGE, SCIENCE, TECHNOLOGY, AND TODAY'S STUDENTS

What makes us [as a species] *different is our language, values, artistic expression, scientific understanding, and technology.*
—Mihaly Csikszentmihalyi (1996)

Language, science, and technology have always provided models for helping us understand the natural world. Setting the stage for helping students comprehend how these subjects relate to understanding and decision making in everyday life means encouraging students to think critically, creatively, and cooperatively. And if you can directly connect some lessons to life outside of school (even as homework), so much the better.

Whether it is approving environmental conservation efforts or reading directions for their stereo, citizens are increasingly called upon to participate in decisions that involve science and language. Poor preparation in the language arts and science frequently result in decisions based more on quirks in the human reasoning process than on informed logic. As John Paulos (1989) has suggested, our inability to communicate intelligently about science results in misinformed governmental policies, confused personal decisions, and an increased susceptibility to all kinds of pseudoscience.

Lack of information is not as big a problem as the lack of knowledge and wisdom. Today's fast-paced technological and communications culture requires us all to go beyond the glut of computer-generated data to think and solve problems related to real-world situations. Increasingly, everyone must understand the basic outlines of language, science, and technology to understand patterns, solve problems, calculate probabilities, and deal with the uncertainty of a constantly changing world. These areas of concern are having an increasing impact on education, the civic process, and American culture in general. At the very time that they need to understand what is going on, many Americans are being distanced from informed participation.

Teachers are caught between disengaged students and a culture that only pays lip service to high academic standards. However, we are beginning to get external structures like national goals and subject-matter standards that give students the signal that hard work and academic achievement will be rewarded. We are far from the goal of being a nation of readers or first in the world in science, as stated in *Goals 2000* (National Governors' Association, 1987). A few positive trends have become evident. In 1995 the Council of Chief State School Officers (CCSSO) released a report that showed positive trends in science education. The report links many of these changes to two factors: the development of unambiguous professional standards and more teachers who really care about these subjects. According to the report, from 1982 to 1994 more students took higher-level science courses. In 1982, only 32 percent of U.S. high school graduates completed three years of high school science, but in 1994 that number had gone up to 51 percent. Similar positive trends were noted at the elementary level (CCSSO, 1994).

There is more good news. The National Assessment of Educational Progress (NAEP) (1995) shows that student scores in science have inched upward over the last ten years. This increase in science achievement is equivalent to achieving one grade level higher. Again, improved student performance is partly attributed to new standards that support active communication, collaboration, critical thinking, and inquiry-based participatory learning. The positive results of this and other programs around the country suggest that there are many ways to go about moving ahead with national standards. There was, however, one constant factor in successful school programs: staff development and supporting teachers in the classroom.

PLANNING FOR INTEGRATED INQUIRY

Integrated inquiry in language and science requires planning and organizing the instructional program so that these disciplines overlap in a manner that relates to everyday experience and the developmental needs of learners. Connecting language and science to the real world of children makes both subjects less abstract and intimidating. Thoughtful real-world questions are usually not constrained within a single discipline. Many can even be framed by circumstances found at home and in individual classrooms. Whether they live in the suburbs or in the city, children are now more likely to grow up without incorporating plants, animals, and other elements of the natural world into their lives. Many of these things can be brought into the classroom in a manner that stimulates student exploration of the natural environment. Central to creating such a positive learning environment is everyone's desire to acquire, share, and construct knowledge.

One way to look at curriculum integration is to view lessons along a continuum. At the lowest stage the teacher would pick a topic, like weather, and study it within specific subjects, sometimes days or weeks apart. In the second stage of subject-matter integration the students are learning about a common topic from at least two subjects that are studied concurrently. At the third and highest level, students and teachers collaborate on a common theme and its contents without worrying about subject-matter boundaries. This is done in a manner that blurs curriculum boundaries within subjects and reaches toward an understanding of the common theme. Some teachers move back and forth between these three levels on an almost daily basis. Whatever the degree of curriculum integration, when it comes to science, language, and technology it works best when tied closely to *inquiry*, the search for information, knowledge, or truth.

In interdisciplinary science, inquiry process skills come in many forms. The basic idea is to have students investigate complex processes in a manner that requires observation, experimentation, analyzing data, and drawing conclusions in order to solve a problem. This way children can use the intellectual tools of science and language and the concrete tools of technology to discover meaningful information, accumulate knowledge, construct meaning around a problem, and propose solutions based on solid reasoning. As students experience, organize, reason, and self-confidently act within a supportive environment, they become capable of using language skills and scientific information to make choices that come up every day.

Measurement, data collection, inquiry across disciplines, and hands-on problem solving are all part of the mix it takes for learners to construct meaning as they interact with each other, the teacher, electronic media, and the curriculum. A major goal is the creation of a high level of interest by involving the children directly and purposely in inquiry. It is clear that the learning process is as important as specific subject-matter content. The teacher's role in interdisciplinary inquiry is often more like that of a knowledgeable guide or advisor who sets the stage. As a resource person or point of reference, the teacher also arranges for imaginative learning materials that encourage children to get to work constructing, experimenting, exploring, and generating puzzling questions to solve.

Understanding the overlapping ideas in science and language involves learning to think critically and learning to create relationships. How these relationships are structured in a student's mind depends on such factors as maturity, physical experience, and social interactions. It is our belief that the ability to inquire, think, collaborate, and investigate will fuel personal autonomy and self-direction in learning.

It is through the beauty of personal and group discovery that one proceeds to self-reliant learning, deep knowledge and freedom.
—Friedrich Schiller (1954)

REFLECTION AND TAKING RESPONSIBILITY

Reflecting is a special kind of thinking that is both active and controlled. When ideas pass aimlessly through a person's mind, or someone tells a story which triggers a memory, that is not reflecting. Reflecting means focusing attention. It means weighing, considering, choosing. Suppose you want to drive home: You get the keys out of your pocket, put them in the car door, and open the door. Getting into your car does not require reflection. But suppose you reached in your pocket and could not find the key. To get into your car requires reflection. You have to think about what you are going to do. You must consider possibilities and imagine alternatives. Let us hope that you have a "hide-a-key" under the car.

Reflection and cooperative strategies can help children take responsibility for their thoughts and provide them with an inner confidence. When the right elements are in place, language, science, and the tools of technology can be used to solve interesting problems in unique ways. As we replace traditional chalk, talk, and textbook methodology, it clearly helps to have each learner reflect on how their reality connects to the subjects being studied. The reasoning skills that stem from literacy allow students to go beyond the scientific method to use language and technological tools to solve problems, discover relationships, and analyze patterns with confidence. Teachers who are enthusiastic about what they are teaching can encourage their students to do this, while instilling a respect for the spirit and beauty of the natural world.

UNDERSTANDING AND EXPLAINING
THE NATURAL WORLD

In language and science teaching there is a complex interplay between communicating meaning, content, and pedagogy. Communication skills go hand in hand with pedagogical and content-area knowledge. Without a reasonably strong content base, teachers will have trouble acting as informed guides for cooperative learning, critical thinking, interdisciplinary inquiry, or anything else. The flip side of the coin suggests that without some knowledge of pedagogy it is difficult to make any subject personally relevant and interesting for students. To be effective, teachers must know both the subject matter being taught and the characteristics of effective instruction.

Throughout the next decade, colleges and universities will prepare teachers to play a crucial role in upgrading linguistic and scientific literacy. They will also be called upon to help school districts with professional development and provide teachers with graduate classes. At all levels of schooling there has long been a breach between what was taught and what was really learned. Understanding and meaningfully explaining the real world—and relating it to personal experience—is different from the interpretations and understandings advanced in most courses of study.

Typical school programs have produced students with increasingly negative attitudes about these subjects as they progress through the grades. This is especially true when language and science courses do not consider the needs, interests, motivations, or experiences of the learners, or when the material being covered is not viewed as useful or valuable (Good & Brophy, 1994). Teaching linguistic and scientific literacy often works best when it is organized around real-life problems. The language instinct is as strong in the human race as is the desire to understand the natural world. People have always been concerned with transmitting attitudes, shared values, and ways of thinking about the natural world to the next generation.

Knowing about science means being able to clarify scientific attitudes and make informed judgments about scientific issues. To do this, students need to cultivate scientific patterns of thinking, logical reasoning, curiosity, an openness to new ideas, and skepticism in evaluating claims and arguments. Understanding certain basic principles of science, being "numerate" in dealing with quantitative matters, thinking critically, measuring accurately, and collaboratively using ordinary tools of science (including calculators and computers) are all important.

To achieve this type of scientific literacy, students need to be able to develop and apply creative and rational thinking abilities. It is important that they hold values and attitudes that promote ethical, collaborative, and moral thinking. Increasingly we expect all students to develop a perspective that promotes the interdependent nature of the environment and global society. Part of scientific literacy is the holistic thinking process across disciplines. This includes the ability to use science concepts, facts, and principles in the solution of problems from many disciplines as they manipulate the materials of science and effectively communicate information.

LANGUAGE AND SCIENCE INQUIRY MODELS

Science and language learning uses a method of scientific inquiry and problem solving that allows students to observe phenomena and understand the realities of the universe; this learning is done collectively, as a cohesive, symbiotic group where ideas and strengths are shared and problems and questions become tools for discovery.

In the latest approaches, language, science, and technology are seen as touching people, caring for the planet, and helping people become knowledgeable and socially responsible citizens. Today's best language and science teaching also emphasizes inquiry and builds on student understandings. A priority is given to improving each student's self-image and self-concept. Both are viewed as an indication of performance. Some of the newest methods include techniques such as creative visualization (guided imagery), keeping daily logs or journals, and expressing attitudes through creative endeavors such as writing, building, or art. Creative and critical thinking is encouraged. As students work on projects and presentations, they combine experiential knowledge with theoretical understandings in language arts and science. Emphasis is on interesting examples, everyday applications, and hands-on collaborative inquiry. We are moving toward a science and language curriculum where the student is viewed as a participant and an explorer—reaching out to acquire the skills they need now and in the future.

Increasingly, school learning is augmented by activities like museum visits, community-group meetings, outdoor education programs, peer teaching, and programs for parents. Even children who live in the city can investigate their urban environment, observing, drawing, and videotaping animals (like squirrels) in "natural" settings. In the classroom, the teacher is becoming as much a facilitator and a learner as an authority. Students are even coming to share in some of the teaching chores. As an interdisciplinary experience, the emphasis is on relating to other subjects and the students' world outside of school. This kind of teaching has implications for teachers. Some suggestions are as follows:

- Integrate learning processes and conceptual knowledge in ways that reflect the richness and complexity of the content.

- Allow time for creativity and incubation of ideas, and encourage students to imagine and question.
- Encourage the use of a variety of materials, technology, and sources, including what might be observed in parks, wetlands, backyards, or schoolyards.
- Use techniques such as brainstorming to generate ideas, open-ended discussions, and collaboration with peers.
- Recognize that what students learn is fundamentally connected with how they learn it.

It is possible to stimulate the exploration of interdisciplinary themes while respecting the separate disciplines and the unique methods for encountering the world. New criteria developed by the National Council of Teachers of English (1996) and the National Science Teachers' Association (NSTA) support their core subjects and the integration of science and language arts. They suggest connecting the following skills: reading, listening, speaking, sorting and classifying, communicating, solving problems, measuring, interpreting data, formulating and interpreting models, and using spatial and time relationships.

Currently, science trade books and published commercial science kits and materials are used as a major source of hands-on science activities and interdisciplinary inquiry. New programs in technology also offer the possibilities for integrating science and language instruction. The technological problem-solving method, for example, finds middle school students building technological solutions to various problems, with relatively little guidance from the teacher. Activities like bridge building are a good way to connect science and communication skills. In the bridge-building exercise, students typically design and build bridges with lightweight materials. These problem-solving activities have proven highly motivating for most students and connect them with more abstract concepts in both science and language (Sanders, 1994).

Bridge-Building Activity

This is an example of an interdisciplinary science and language activity which reinforces skills of communication, group process, language arts, science, and technology.

Materials:
> A lot of newspaper and masking tape, one heavy rock, and one cardboard box. Have students bring in stacks of newspaper. You need about a one-foot pile of newspapers per small group. At least two or three rolls of masking tape is also required for each group of four or five. Bridges are a tribute to technological efforts, and employ community planning, good communication skills, engineering efficiency, mathematical precision, aesthetics, group effort, and construction expertise.

Procedures:
1. For the first part of this activity, divide students into groups of about four. Each group will be responsible for investigating one aspect of bridge building.

   Group One: Research

   This group is responsible for going to the library and looking up facts about bridges, collecting pictures of kinds of bridges, and bringing back information to be shared with the class. Use the Internet if possible.

   Group Two: Aesthetics, Art, Literature

   This group must discover songs, books, paintings, artwork, and so on which deal with bridges.

   Group Three: Measurement, Engineering

   This group must discover design techniques, blueprints, angles, and measurements of actual bridge designs. If possible, visit a local bridge to look at the structural design, or a similar activity. Each group presents their findings to the class.

2. Have the group representatives get together to present their findings to the class. Allow time for questions and discussion.

The second part of this activity involves actual bridge construction by the students. When there is only a fifty-minute time frame, we start with point 3.

3. Assemble the collected stacks of newspaper, the tape, the rock, and the box at the front of the room. Divide the class into groups. Each group is instructed to take a newspaper pile to their group and several rolls of masking tape. Explain that the group will be responsible for building a stand-alone bridge using only the newspapers and tape. The bridge is to be constructed so that it will support the large rock and so that the box can pass underneath. Planning is crucial. A group can talk and plan together for ten minutes. Explain that during the building time students will not be allowed to talk. Be sure to tell them that they only have fifteen minutes to do the actual building. When finished, they can choose which of the two tests they want first.

4. Each group is given ten minutes of planning time in which they are allowed to talk and plan together. During the planning time they are not allowed to touch the newspapers and tape, but they are encouraged to pick up the rock, make estimates of how high the box is, make a sketch of the bridge, or assign group roles of responsibility.

5. At the end of the planning time, students are given about fifteen minutes to build their bridge. During this time there is no talking among the group members. They may no longer handle the rock or the box: It is newspaper and masking tape time. A few more minutes may be necessary to ensure that all groups have a chance of getting their constructions to meet at least one of the two tests (rock or box). If a group finishes early, they can add some artistic flourishes to their bridge or watch the building process

in other groups. (Remember, you do not want every group to pass both tests. If pressed for time, you can simply end with testing each bridge.)

Evaluation:
> Stop all groups after the allotted time. Survey the bridges with the class and allow each group to try to pass the two tests for their bridge. Does the bridge support the rock and does the box fit underneath? Discuss the design of each bridge and how they compare to the bridges researched earlier. Students may wish to create awards for the "best" bridges. Awards could be given for the most creative bridge design, the most sturdy bridge, the tallest, the widest, the best cooperative group, and so on. You might want to take photographs of the process.
>
> Students may also evaluate their teamwork and communication. Without a great deal of both of these, no tests will be passed.

Follow-up/Enrichment:
> As a follow-up activity, have each group measure their bridge and design a blueprint (include angles, length, and width of the bridge) so that another group could build the bridge by following this model.

TURNING TOWARD COMPLEMENTARY POSSIBILITIES

Inadequate language skills and no opportunity for scientific inquiry are cited as a major reasons for students failing in school (American Association for the Advancement of Science, 1990). Some of the failure is due to students being disengaged and unmotivated in school. Some is due to inferior social status—less health care, housing, out-of-school preparation, and so on. And some is due to a tradition of teaching that does not match the way students learn. Active inquiry, cooperative teams, and interdisciplinary reasoning are all paths to motivation and a richer curriculum.

The communicative instinct to speak, learn, and understand nature is so tied in with human experience that we have trouble picturing what life would be like without it. Even when working in cooperative groups, students construct knowledge based on their own personal experience. Language and science instruction now aims to help students learn how to apply knowledge, solve problems, and understand concepts. Students need to be able to use the skills of language and science processes to change their own theories and beliefs in ways that are personally meaningful. By constructing their own knowledge in a meaningful context, children can gain a conceptual understanding and develop the means for integrating language and science knowledge into their personal conceptions. Even young children can be encouraged to creatively pursue their own investigations.

To really learn the skills of language and science students must follow a learning cycle: explore new phenomena, construct their own understandings, examine, represent, solve, transform, apply, prove, and communicate. This involves discussing their work with peers, being responsible for each other, and accepting a level of individual accountability. When students work together in groups, they can discuss, pose questions, analyze, create theories, research on the Internet, and make presentations. This gives them the social support they need to reach across subject-matter boundaries and move toward unexplored possibilities.

Tapping into a child's natural curiosity with thematic units is a powerful way of using the intellectual tools of science, technology, and the language arts. Nourishing the big ideas of the natural world (science) requires increasingly sophisticated levels of language usage. New information technology has a lot to do with this. Computers change the dynamics of communication and the teaching of science. As technology allows us to step into stories for the first time, it is bound to effect how we view literature, learning, and ourselves. As students can read, hear, experience, and direct visual imagery at the same time, interesting new stories will be constructed and forbidding material will be brought to life in an engaging way.

As the national standards project for the language arts has pointed out, literacy in today's world must extend beyond reading and writing to include oral communication, visual texts, technological displays, and interdisciplinary inquiry. The National Commission for Excellence in Teaching extends this to communicating collaboratively, adapting to changing situations, thinking critically, acting imaginatively, and dealing with a world filled with the products of scientific inquiry.

What is most important for interdisciplinary inquiry with science and the language arts? *Highly skilled teachers.* Because of their daily work in the classroom, teachers are key to helping students deal with the expanding definitions of literacy. They must be enthusiastic, up to date, and able to use the whole range of communication tools to get at the power and beauty of scientific understanding.

> *A good teacher can provide astonishing revelations. Good teachers put snags in the river of children passing by and, over the years, they redirect hundreds of lives.*
> —Tracy Kidder (1989)

Chapter 2

Promoting Critical and Creative Thinking

> *Critical literacy is the ability to move beyond literal meanings, to interpret texts and to use writings not only to record fact but also to analyze, interpret, and explain.*
> —Morriss & Tchudi (1996, p. 12)

Creative and critical thinking are an intragal part of today's elevated concept of literacy. No wonder that fostering the critical and creative imagination is one of the most important recommendations found in the new science and language arts standards. This chapter explores the possibilities for thoughtfulness in the inquiry-based science and integrated language-arts classroom. The emphasis is on thinking, not telling children what to think.

How can we be more efficient at sifting out truth from opinion and foolishness when using today's wide range of communications tools? From astrology to the Roswell aliens, pseudoscience is found on toystore shelves, the Internet, and the mass media. The antidote? A supportive community of responsible, skeptical, and reflective learners. A sound grounding in science also helps. The use of empirical standards, logical arguments, and skepticism is how 20th-century science has distinguished itself from other disciplines or ways of knowing. Solid thinking and science process skills can help us tell reality from fantasy and science from superstition. The mass media could make a contribution by encouraging a few more television or movie portrayals of ethical practitioners of the scientific method.

Students need to know what science and technology can and cannot contribute to society. New curriculum standards recognize this and go on to suggest paying more attention to the world outside of school. One way to encourage this connection is to encourage thinking that is similar to that used by scientists as they go about designing real-world applications. Creative thinking and skeptical inquiry flow from the imagination. Scientists collect data, select information, and reflect on what it might mean for the natural world. As students use thinking skills to solve problems within such a framework, the more creative may approach a task in an unconventional, spontaneous, flexible, and original manner. Sometimes this is done within a preexisting paradigm, sometimes it means breaking out of conventional boundaries.

Many of the influences on thinking are generated far from the scientific and educational world. Children who grow up in a nonlinear world of television, computers, and the Internet may have a jump on grown-ups in adapting to the chaotic world of technological change. Teachers may have to learn about some of this from their students, because the 21st century will place a higher value on sorting through the information glut and doing many things at once.

CREATIVE AND CRITICAL THINKING

The national standards movement of the 1990s has supported the movement toward an inquiry-based interdisciplinary curriculum. This has gathered momentum based on the research, the literature, and what worked in science and language-arts instruction. When combined with energized teachers, the results were lessons that excited children and helped them become intellectually productive. Like the ideas presented here, the standards make suggestions and encourage teachers to construct their own understanding how thinking skills can play a more central role in their classroom.

Creative and critical and thinking are constructed by the mind and built on personal experience. Various attributes of thoughtfulness develop along with other elements of functional intelligence and personality. The quality of an individual's thinking is influenced by experiences, culture, emotions, home environment, and educational possibilities. As children learn language communication skills and science-process skills, they grow better at reasoning inductively and deductivity.

Being good at thinking means being able to form alternative explanations and demonstrate intellectual curiosity in a manner that is flexible, elaborate, and novel to the thinker. As part of their responsibility to the future, teachers must respect the unique ideas developed by children and encourage creative and critical thinking. It seems clear that many future problems will be solved by people who are flexible, open, original, and creatively productive. So what students can actually do with scientific and communications knowledge is of utmost importance.

Good critical-thinking activities encourage students to analyze underlying assumptions that influence meanings and interpretations of information. Such intellectually demanding thinking leads children to identify, clarify, problem solve, and become more productive creatively. The questions explored can be as general as "Are there limits to how much of the physical universe we can understand?" and "How secure are the foundations of knowledge in science and language arts?" Questions can be "How did you figure that one out?" or "What does it mean?" The wording may be changed, but children are never too young to analyze the underlying assumptions that influence meanings. And they are never too young to question the interpretation of findings and participate in the act of knowledge creation.

We undermine our chances for creativity by getting caught up in what David Whyte (1994) calls "the eddies and swirls of everyday existence." He makes this clear when he cites a line of poetry from one of his adult students:

> I turned my face for a moment
> and it became my life. (p. 23)

Though we often connect the characteristics of critical and creative thinking, there are traditional differences between the two. Here, *critical thinking* is viewed as reasoning, criticism, logical analysis, searching for supporting evidence, and evaluating outcomes. Activities that support this involve clarifying problems, considering alternatives, strategic planning, problem solving, and analyzing the results. *Creative thinking* is viewed as fluency, flexibility, originality, and elaboration. We would hope that skills developed in this area would result in the creation of unique expressions, original conceptions, novel approaches, and demonstrations of the ability to see things in imaginative and unusual ways. Problem solving and implementation are part of both creative and critical thinking. Here we are less concerned with definitions than with encouraging children to develop skills that lead to high-quality thinking.

THINKING ACROSS SUBJECTS AND IN EVERYDAY LIFE

One way to look at modes of thought across disciplines is through *symbolic, imagic,* and *affective* thinking. Symbolic includes using words, numbers, and other symbol systems. Imagic is visual, spatial, tonal, and kinesthetic. It involves the kind of imagery used by scientists and architects, sound relationships explored by musicians, and the movement found in sports and dance. Affective thinking works with emotions and feeling to direct inquiry. All three modes of thinking build on reasoning and intuiting to connect the analytic to the intuitive.

The power of interdisciplinary science and language-arts inquiry lies in its possibility for building on alternative ways of knowing to encourage diversity of thought and increase the chances of making creative connections. Creativity might be changing how a particular subject is studied or changing some of the elements of personal life. At the highest level, a creative mathematician might change the way science is applied to scientific problems. In our personal lives, creativity could mean changing day-to-day practices to allow for an hour of exercise to improve the general quality of life. An example that is more relevant to children at school would be clever hypothesis formation. Effective instruction in science and language arts provokes students to create their own questions and think of creative applications in the world outside of school. As children become interested in such intellectual invention, it is important for the teacher to hold off on judgments and let the evidence itself be the judge.

Scientists use the tools of science and language to collect, examine, and think about data. Conclusions are formulated and outcomes explained. Like scientists, students can reason, analyze, criticize, and advocate, while avoiding dangerous materials and problems that are developmentally inappropriate. They can also learn to think spontaneously, flexibly, and originally. An understanding of the physical and biological universe is most solid when it builds on a child's own experiences and discoveries. By modeling thoughtful behavior, teachers can help students become self-confident enough to resolve inconsistencies and uncover truths in science and language arts.

Developing methods that extend critical thinking across the curriculum is a reform strategy that is widely supported and working. Teaching thinking skills involves developing the ability to assess information and make creative and critical judgments. Critical and creative thinking have shown staying power and a dispersion of expertise. Some teachers integrate thinking skills into each subject in the curriculum. Others directly teach children thinking skills and strategies. A third approach uses metacognitive (thinking about thinking) strategies. Another conceptual framework synthesizes all three and adds the heavy use of visual images to soften the boundaries of subject matter and encourage thinking across disciplines. Of course, many teachers pragmatically borrow from all the available possibilities to tailor their own lessons for interdisciplinary inquiry in science and language arts.

MANY WAYS TO ACCESS KNOWLEDGE

Recognizing thinking skills as directly involved in successful learning throughout the curriculum does not come as a surprise to most teachers. There is, however, a tendency to think of science and language as clear and clean; in science you formulate hypotheses, organize experiments, collect data, analyze, and interpret the findings. Linguists are concerned with the system of sounds of language; its sentence structure, spelling, and so on. As scientists, communications experts, and engineers who are doing original work will tell you, the reality is far less clear-cut and tidy. There are many false starts and detours as they work through alternatives to discover relationships and invent new perspectives. What makes it satisfying for many scientists is the sheer power of searching at the frontiers of knowledge. This passion for inquiry and feeling outward into space for new experiences is just as important for children.

Creative and critical thinkers tend to be reflective, think problems through, and have a flexibility to consider original solutions and a curiosity to pose and expand new questions. The research evidence suggests that giving students multiple perspectives and entry points into subject matter increases thinking and learning (Sears & Marshall, 1990). The implication here is that ideas about how students learn a subject need to be pluralized. Almost any important concept can be approached from multiple directions, emphasizing understanding and making meaningful connections across subjects. This means making available learning possibilities and resources (human and technological) that might appeal to pupils with very different learning styles and cultural backgrounds.

Tomorrow's schools will need to incorporate frameworks for learning that build on the multiple ways of thinking and representing knowledge. By organizing lessons that respect multiple entry points to knowledge, teachers can enhance thoughtfulness and make the school a home for inquiry. If many of today's dreams, possibilities, and admired models are going to be put into widespread practice, we all must be more courageous in helping move good practice from the margins into the schools, the media, and the home.

A child's thinking ability evolves through a dynamic of personal abilities, social values, academic subjects, and out-of-school experiences. We are all involved, directly or indirectly, in the education of children. Revitalizing the educational process means recognizing the incomplete models of how the world works that children bring to school with them. From birth, children are busy making sense of their environment. They do this by curiously grappling with the confusing, learning ways of understanding, developing schemes for thinking, and finding meaning. As they enter school, children can sing songs, tell stories, and use their own processes of reasoning and intuiting to understand their surroundings. By the time they reach first grade, they have already developed a rich body of knowledge about the world around them. The best beginning can be extended in school when the teacher cultivates a broad disposition to critical thinking throughout the year. Working with natural rhythms is important, but it takes learning-centered instruction to continue the process of developing mature thinkers.

MERGING THINKING WITH CONTENT AND EXPERIENCE

Individual reality can be solidly constructed with real-life experiences that evoke personal meaning in the learner. Science and language-arts lessons may begin with real materials, invite interactive learning, and allow children to explore the various dimensions of thoughtfulness, subject matter, and real-world applications. The goal is to help children construct a new set of expectations and establish a new state of understanding.

When students make sense of something by connecting to a set of personal everyday experiences, constructivists may call it "viable knowledge." Whether they are familiar with the terms being tossed about, good teachers have always connected academic goals to practical problem solving and students' life experiences. Using such a real-world base embeds thinking skills into the curriculum so that students are intensely involved in reasoning, elaboration, hypothesis formation, and problem solving. Such inquiry-based learning cannot be isolated within calcified disciplinary boundaries.

Developing mature thinkers who are able to acquire and use knowledge means educating minds rather than training memories. Sometimes the acquisition of enhanced thinking skills can be well structured and planned, at other times it is a chance encounter formed by a crazy collision of elements. The ability to raise powerful interdisciplinary questions about what is being read, viewed, or heard is a dimension of thinking that makes a powerful contribution to the construction of meaning. When motivated to reason intelligently, children come up with good decision making and elaboration. Out of this comes insightful creations that suggest possibilities for action. As all of these elements come together, they form the core of effective thinking and learning.

NEW APPROACHES TO THINKING AND PROFESSIONAL DEVELOPMENT

Implementing new methods for teaching science and language-arts instruction depends on teachers who purposely invite reflective thinking.

This means that both perspective and practicing teachers must take science and language-arts courses in which they learn science through inquiry and learn to apply language-arts concepts within a context similar to the one they will arrange for their students. When carried out over time, professional development activities have proven useful in helping teachers organize instruction to accommodate new ways of representing and imparting knowledge. The results expand horizons and organizational possibilities. When teachers actually *do* interdisciplinary inquiry they can reflect on teaching practice with colleagues and add to their ever-evolving base of good instructional practice.

Creative and critical thinking are natural human processes that can be amplified by awareness and practice. Both creative and critical thinking make use of specific core thinking skills. Classroom instruction and guided practice in the development of these skills will include the following:

1. The ability to attend to selected chunks of information—this includes skills such as defining, identifying key concepts, recognizing the problem, and setting goals.
2. The ability to recognize important content information—skills such as observing, obtaining information, forming questions, and clarifying through inquiry are some of the skills that fall under this category.
3. The ability to remember—encoding and recalling are thinking skills which have been found to improve retention. These skills involve strategies such as rehearsal, mnemonics, visualization, and retrieval.
4. The ability to organize and arrange information so that it can be understood or presented more effectively—some of these organizing skills consist of comparing, classifying (categorizing), ordering, and representing information.
5. The ability to analyze—analysis is at the heart of creative and critical thinking. Analysis includes recognizing and articulating attributes and component parts, focusing on details and structure, identifying relationships and patterns, grasping the main idea, and finding errors.
6. The ability to use prior knowledge to add information beyond what is known or given—connecting new ideas, inferring, identifying similarities and differences, predicting, and elaborating adds new meaning to information. This involves such higher-order thinking as making comparisons, constructing metaphors, producing analogies, providing explanations, and forming mental models.
7. The ability to integrate information—integrating involves putting things together, solving, understanding, forming principles, and composing and communicating. These thinking strategies involve summarizing, combining information, deleting unnecessary material, graphically organizing, outlining, and restructuring to incorporate new information.
8. The ability to assess and evaluate the reasonableness and quality of ideas—skills of evaluation include establishing criteria and proving or verifying data (Marzano et al., 1988).

For teachers to build a solid base of thinking skills into daily science and language-arts lessons requires that they consciously question and reflect on the best approach. Introspective questions about the characteristics of effective instruction help. For example, "How can I get students to focus their thinking, ask questions, retrieve new information, and generate new ideas for analysis?" is a good start.

The Think-Aloud Questioning Technique

Students are asked to describe their thinking as they do activities, experiment, and work through problems. The teacher asks children to talk their way through whatever they are doing and follows up with a few open-ended questions: for example, "Tell me what you are thinking," "Can you tell us more?" "Why?" "And, so . . . " (asking the student to end the comment). Think-alouds have been shown to help teachers informally assess student thinking and concept development in the language arts and science. They have also proven effective in helping students clarify their own thinking and assume ownership of the process (Wade, 1990). Often children come up with something the teacher may never have thought of. There are no "wrong answers" as students describe their thought processes.

MULTIPLE WAYS OF THINKING AND APPLYING INFORMATION

The real voyage of discovery consists not in seeking new landscapes, but in having new eyes.
—Marcel Proust (1981, 1:32)

Beyond using investigations in science and literature-based reading in language arts, teachers are bringing these subjects to life by setting thoughtful application problems in real-life contexts. Knowledge is particularly useful when it can be applied or used to create new knowledge. Students need opportunities to use their knowledge to compose, make decisions, solve problems, and conduct research to discover more. As teachers facilitate activities built on multiple ways of reasoning, doors are opened to the physical and biological universe.

The infusion of creative thinking into the science and language-arts curriculum goes hand in hand with the basic principles students must learn to be competent in these subjects. Solid reasoning supports the foundation of interdisciplinary inquiry, real-world applications, and the production of new knowledge. In our efforts to bring science and language arts to life by making them relevant to students' daily lives, it is important to leave spaces where students and teachers can reflect on what they are doing and figure out where they will use the skills that they are learning. Creativity has as much to say to government, business, and education as it does to creative fields like the arts and sciences.

People who think creatively have the ability to produce and consider many alternatives, creating or elaborating on original ideas. Creative thinkers have the ability to see multiple solutions. Developing and expressing emotional awareness is a part of creative thinking. This is frequently done by perceiving and creating images which are vivid, strong, and alive, both from an internal and an external vantage point. Making use of imagination, movement, and sound in playful and useful ways is another element of creative thinking. Overcoming limitations and creating new solutions, using humor, predicting consequences, and planning ahead are other elements.

Students will learn elements of creative thinking from interpersonal communication behaviors. These are developed in a variety of ways: listening, speaking, arguing, problem solving, clarifying, and creating (Dissanayake, 1992). Pairs of students can argue an issue with other pairs and then switch sides. The chaos and dissonance of group work can help foster thinking and imaginative language development. This way students learn to work creatively with conflicts, viewing them as possibilities for the improvement of literacy. Hopefully some of this will carry over to conflict resolution and peer resolution of other disputes.

Thinking does not thrive in a threatening, intimidating environment where either adult or peer pressure impedes independence. A classroom organized for creative science and language-arts group work function as a community that respects individual learners. Good teachers support diverse thinking styles and collaboration, helping all students to think and step outside of subject matter and experience boundaries to construct meaning. This means that the teacher and students open themselves up to suggestions, styles of thinking, connections, and ambiguities previously unexamined. The potential for imaginative action grows out of this process. As Aristotle suggested, there are two steps to doing anything: Make up your mind, and do it.

PROMOTING CRITICAL THINKING

For the 21st century, the ability to do a whole host of complicated tasks at the same time (multitasking) may prove crucial. For those who have trouble even walking and chewing gum at the same time, there is trouble ahead. The multidimensional search for meaning is made at least a little easier when there is a supportive group climate for generating questions and investigating possibilities. Critical-thinking questions may also come into play after solutions are put forward. Ask students to analyze problems they have solved: As they examine how underlying assumptions influence interpretations, children can be pulled more deeply into a topic, and by evaluating their findings on the basis of logic they find other possibilities.

The following are useful stimulants to critical thinking and interdisciplinary inquiry:

- Provide opportunities for students to explore different viewpoints and domains of information that arouse frustration or outrage.
- Conduct debates and discussions on controversial issues that somehow connect to science, language arts, and technology. Students work in groups to present an argument on a topic and present their view to another group. Sides can then be switched, the opposite view defended, and different routes to a better social order explored.
- Role play historical events or current news happenings from conflicting viewpoints. Examine some of the more questionable television news images, whose power is palpable but whose connection to reality is tenuous.
- Watch TV broadcasts that present different viewpoints (for example, those that interview individuals with differing perspectives on a problem).
- Have students write letters to the editor of newspapers, TV, or popular journals expressing their stance on an issue of importance.

These suggestions open up the possibility for developing thinking by practicing argumentative thinking skills. The basic goal is to stimulate and encourage a wide range of collaboration, divergent thinking, and discussion. By arguing important moral dilemmas in science, medicine, technology, politics, literature, art, music, or sports, students can learn content, reasoning possibilities, and extended ethical concepts. To have power over the story that dominates one's life in these technologically intensive times means having the power to retell it, deconstruct it, joke about it, and change it as times change. Without this power, it is more difficult to think and act on new thoughts and open the doors to deep thinking.

Beyond specific teaching strategies, the climate of the classroom and the behavior of the teacher is very important. Teachers need to model critical-thinking behaviors, setting the tone, atmosphere, and environment for learning. Being able to collaborate with other teachers can make a formative contribution to how the teacher might better see and construct individual classroom reality. In collaborative problem solving, teachers can help each other in the clarification of goals. They also share the products of their joint imaginations. Thus, perceptions are changed, ideas flow, and practice can be meaningfully strengthened, deepened, and extended. Like their students, teachers can become active constructors of knowledge. The essence of creative thinking is questioning the governing realities of the institutions in which we find ourselves.

SELF-RELIANT THINKING

The old view of teaching as the transmission of content has been expanded to include new intellectual tools and new ways of helping students thoughtfully construct knowledge on their own and with peers. Teachers who invite thoughtfulness understand that knowledge is to be shared or developed rather than held by the authority. They arrange science and language-arts instruction so that children construct science and language concepts and develop their thinking skills. As a result, everyone involved becomes an active constructor of knowledge and more capable of making thoughtful decisions in the future.

Recognizing the development of thinking skills is a good first step toward application and assessment. Some possible guideposts for assessing development of self-reliant thinking and collaboration include the following:

- Avoid "How do I do it?" questions (students should ask other group members before asking the teacher).
- Use trial-and-error discovery learning.
- Ask powerful *why* questions when questioning peers and teachers.
- Use metaphor, simile, and allegory in speaking, writing, and thinking.
- Develop interpersonal discussion skills for shared inquiry.
- Increase the ability to work collaboratively.
- Show willingness to begin a task.
- Initiate inquiry.
- Increase comfort with ambiguity and open-ended assignments.
- Synthesize and combine diverse ideas.

It is hard to measure attitudes, thinking, and interpersonal skills on a paper-and-pencil test. Another way is to observe the humor, antidotes, parental reactions, and teacher–student interaction. The ability of both students and teachers to pull together as a team influences how well students reflect on their thinking, pose powerful questions, and connect diverse ideas. Failure to cultivate these aspects of thinking may be a major source of difficulty when it comes to learning content (Anderson & Burns, 1989).

POSSIBILITIES FOR ILLUMINATING LEARNING

A curriculum that ignores the powerful ideas of its charges will miss many opportunities for illuminating the human condition. To teach content without regard for self-connected thinking prevents subject-matter knowledge from being transformed in the student's mind. If the curriculum is to be viewed as enhancing being and opening to the unfamiliar, rather than merely imparting knowledge and skills, then reasoned decision making is part of the process (Berman, Hultgam, Lee, Rivkin, & Roderick, 1991). A curriculum that takes students' thinking seriously is more likely to be successful in cultivating thoughtfulness and subject-matter competence. Respecting unique thought patterns can also be viewed as a commitment to caring communication and openness.

Breaking out of established patterns can be done collectively or individually by those most directly involved. All of us need the occasional push or encouragement to get out of a rut. It is important for teachers to develop their own reflection and inquiry skills so that they can become students of their own thinking. When a teacher decides to participate with students in learning to think on a daily basis, they nourish human possibilities. Can teachers make a difference? Absolutely. The idea is to connect willing teachers with innovative methods and materials so that they can build learning environments that are sensitive to students' growing abilities to think for themselves. By promoting thoughtful learning across the full spectrum of personalities and ways of knowing, teachers can make a tremendous difference and perform a unique service for the future.

When the ideal and the actual are linked, the result can produce a dynamic, productive, and resilient form of learning. What we know about teaching and thinking is increasingly being put into practice in model classrooms and schools (Boomer, 1988). These exemplary programs recognize that powerful inquiry can help students make personal discoveries that change thinking, and that critical thinking skills can turn an unexamined belief into a reasoned one. Thus, the control people exercise over their lives can be enhanced by the teacher's support of inquiry and caring.

SOME INTERDISCIPLINARY ACTIVITIES FOR TEACHERS

By nurturing informed thinking and awareness, teachers help students learn how to actively apply knowledge, solve problems, and enhance conceptual understanding. As children use these processes to change their own theories and beliefs, they grow in ways that are personally meaningful. In developing conceptual understanding by looking at science and language arts from new angles, students integrate content and thinking skills into their personal experience.

Understanding the essence of the argument means understanding some of the universal truths that speak to everyone and recognizing how a diversity of new voices can add vigor to learning. As students learn about the perspectives of other cultures—including social and historical background—they can explore where stereotypes come from. This should not stand in the way of supplying students with the common and universal roots of present conditions. It means a more integrative understanding of human community and an appreciation of overlapping cultural experiences. It also means including active learning techniques, where students can collaboratively shape alliances that view each students cultural background as a valuable tool for learning and a vista from which they may see and hear anew.

MULTIPLE INTELLIGENCES

Learning has a lot to do with finding your own gifts. To make learning more accessible to children means respecting multiple ways of making meaning. The brain has a multiplicity of functions and voices that speak independently and distinctly for different individuals. Howard Gardner's (1993) framework for multiple entry points to knowledge has had a powerful influence on science and language-arts standards. The standards clearly recognize alternate paths to learning and provide a framework for thinking about science and language arts that is built on recognizing the uniqueness of each child.

Gardner's multiple intelligences are as follows:

1. Verbal–linguistic—ability to use language to excite, convince, convey information.
2. Logical–mathematical—ability to explore patterns and relationships by manipulating objects or symbols in an orderly manner.
3. Musical–rhythmic—ability to perform, compose, or enjoy a musical piece.
4. Spatial–visual—the ability to understand and mentally manipulate a form or object in a visual or spatial display.
5. Bodily–kinesthetic—the ability to use motor skills in sports, performing arts, or art productions.
6. Interpersonal—the ability to get along with others.
7. Intrapersonal—the ability to understand one's inner feelings, dreams, and ideas.
8. Naturalist—the ability to recognize and categorize plants, minerals, and animals; sensitivity to the natural world (Gardner, 1993).

It is possible to take issue with Gardner's interesting approach on several points, such as not fully addressing spiritual and artistic modes of thought. But there is general agreement on a central point: Intelligence is not a single capacity that every human being possesses to a greater or lesser extent. There are multiple ways of knowing and learning. Consequently, methods of instruction should reflect different ways of knowing.

Will these "intelligences" be as central to the 21st century as they were to the 20th century? Does this conception of multiple paths to knowledge adequately cover spiritual and artistic intelligences? With some humor, Gardner recently agreed to give a point for sensitivity to the natural world and half a point for the spiritual. It is important to note that he is not as infatuated with his ideas as are many educators. But we all agree that the lens he provides gives us an interesting way to inquire about critical and creative thinking. And unlike the usual vague psychological theory, this one has direct implications for teaching. Working out the ecology of thoughtfulness requires taking risks with such bold and explicit insights.

CREATIVE AND CRITICAL THINKING ARE PART OF ANY FUTURE

Forecasting the specifics of the future is an essentially harmless form of infotainment. Go ahead, indulge yourself; just do not confuse the fun with what is really going to happen. The more precise the prediction, the more likely it is to be off the mark. Since it is so difficult to figure out what knowledge will be crucial to students in the future, it makes sense to pay more attention to the intellectual tools that will be required in any future. This suggests focusing on how models of critical thought can be used differently, at different times, and in different situations. The idea is to put more emphasis on concepts with high generalizability, like collaborative problem solving, reflection, perceptive thinking, self-direction, and the motivation needed for life-long learning.

Information is not a substitute for thinking, but information and thinking are not antithetical. Higher-level thinking requires quickly sorting through a wealth of information to be effective. There will never be enough time to teach all the information that we feel is useful. But time must be taken to be sure that student thinking can transform knowledge in a way that makes it transferable to the outside world (Sigler, 1985). When there is time for inquiry and reflection, covering less can actually help students learn more deeply.

Topical knowledge (content facts), procedural knowledge (how to study and learn), and self-knowledge are all part of critical thinking. All of these thinking skills are learned through interaction with the environment, the media, peers, and the school curriculum. Some students pick it up naturally, while others learn reasoning skills with difficulty or not at all. Everyone can do it. It is clear that the following thinking skills can be taught directly: generating multiple ideas about a topic, summarizing, figuring out meaning from context, understanding analogy, and detecting reasoning fallacies (Segal, Chipman, & Glaser, 1985).

For students to learn how to think new thoughts requires creative teachers who recognize the need of children to learn in meaning-centered explorations. Feeling and meaning can be turned inside out as children learn how to construct their own knowledge and absorb new experiences in ways that make sense to them. The basic idea here is to go beyond giving students the truth of others, making it possible for them to discover their own.

New ways of considering literacy require teachers who are free to invent diverse approaches to helping students succeed. Expressing the results of reasoning in science and language can take many forms. For example, painting, music, and dance can be resonant with thoughtful meaning. So can harnessing the spirit of inquiry by focusing on questions that can be answered by using the processes of scientific reasoning. The basic idea is to have students inquire with various intellectual tools and media so that they can go beyond the literal and linear to probe areas that are ambiguous in meaning and rich in allusion.

Children can demonstrate their reasoning ability in a number of ways: think-outlouds, videos, performances, photo-college, stories for the newspaper, websites on the Internet, or multimedia projects that can be shared with other students and members of the community. We are already seeing glimmers of a computer-based medium that is broadly expressive and capable of capturing many aspects of human consciousness. As we move into the 21st century, it is bound to reshape the whole spectrum of expression. Communication and information technology sometimes complement and sometimes supersede previous media. Still, the basic learning process and the essence of any curriculum will continue to involve ways of engaging students in thought that matters and sharing what they find.

When knowledge means something to students and they have the skills to make it do something for them, they can do more than learn: They can change the shape of ideas. Critical and creative expression is not limited to print, writing, or electronic imagery. As Isadora Duncan, a famous dancer, is said to have remarked, "If I could write it, I wouldn't need to dance it." To be educated means anticipating and exploring (from many angles) the depths that wait for us under the surface of things, whatever those things may be.

Chapter 3

Cooperative Learning: Teamwork and Individual Accountability

Thought development is determined by linguistic tools of thought and by sociocultural experiences.... The child's intellectual growth is contingent on mastering the social means of thought and language.
—Vygotsky (1962)

Cooperative learning moves students from working alone to working in small groups where they take responsibility for each other. Though there are provisions for individual accountability, students receive information and feedback from peers as well as their teacher. In the cooperative classroom, students become a community of learners working together to enhance individual knowledge, proficiency, and enjoyment. Collaborative small-group learning has become popular because of its potential for motivating and academically engaging students within a social setting.

The essential elements for cooperative learning include positive interdependence, face-to-face interaction, personal responsibility, interpersonal skill development, and group processing. Science and language-arts instruction can be connected and enhanced with cooperative learning. Today's technologies and the technologies of the future can be powerful allies in promoting social interaction and learning. The use of computers, electronic networks, and multimedia technologies in the classroom have at least the potential for helping with the cognitive and social processes involved in a broad range of literacy practices.

COOPERATIVE LEARNING: A SHARED LEARNING ADVENTURE

Cooperative learning involves working together to accomplish shared goals that are beneficial to individuals and the group. Students are able to learn together and perform alone within an environment that allows them to actively construct knowledge. In the cooperative school, communal responsibility and civic engagement are not viewed as optional extras. Everyone is involved and cooperation becomes part of the fabric of schooling at every level. Building team-based organizational structures in the classroom can even make it easier for teachers to reach out to their peers and ensure that colleagues are successful.

A common element in successful schools is a shared sense of community and a socially integrating sense of purpose. Shared common interests and common ground make for better civil discussions than in a school or community culture that places an absolute value on individual choice and unconditional rights. Schools, like the community in general, need common spaces where people meet and share a life in common, a natural fit with cooperative learning. When there is no limit on individual self-realization, public spaces decay and civic culture is weakened. As teachers develop their skills with cooperative learning they can foster quality common spaces and encourage civic conversation across group lines. The whole process can even contribute to collegial teams and help establish a true learning community for teachers.

Cooperative learning is one of the success stories of educational reform. It has solid teacher support and a favorable research base. Researchers have also commented on how cooperative learning improves problem-solving skills and attitudes toward science and language-arts learning (Tobin, 1990). More and more teachers are incorporating the concept of group work (with individual accountability) into their classrooms. It is also becoming common to see suggestions for active-learning teams in new content standards and classroom designs.

Cooperative learning is more than having students cooperate in a group activity or project. Cooperative learning groups employ a set of strategies to encourage having students cooperate while learning in a variety of settings and disciplines at different grade levels. The process involves promoting positive interdependence by dividing the workload, providing joint rewards, holding individuals accountable, and getting students actively involved in helping each other master the topic being studied.

CONNECTING STANDARDS TO THE COLLABORATIVE CLASSROOM

Underlying the more general "opportunity to learn" and "delivery" standards of *Goals 2000* (National Governors' Association, 1987) legislation is the concept of students working together to gain a powerful understanding of how strands of content can be integrated and synthesized. Unlike some earlier innovations, cooperative learning has shown real staying power and has been incorporated into content standards that spell out the knowledge and skills that students should have and performance standards that suggest actual performance levels that students should reach.

Cooperative learning has permeated the standards projects and teacher training from the preservice to the inservice level. The National Science Teachers' Association (1997) and the NCTE (1996) have standards which reflect these principles: science and language is for all students, learning is an active process, science and language arts are based in inquiry, and communicating and working with others is an integral part of the knowledge process. All these principles leave space for collaboration. In addition, cooperative learning is one of several approaches and methods that have been consistently supported by research on science and language-arts teaching (Steinbrink & Stahl, 1994). As teachers learn when and how to structure instruction cooperatively, the process transforms itself from a hot new method into a routine part of instruction.

All too often, neither teachers nor students examine how cooperation plays an important role in their community and in their country. Whether it is the role of cooperation between police and fire departments or between Congress and the president, successful policies require a mix of sticking to your beliefs, cooperation, compromise, and mutual support. One side does not have all the answers. When cooperation is diminished, the political dialogue can become debased, resulting in public anxiety, a frustration about the future, and a scarcity of new ideas. There are two notes of optimism: When expectations are low it is relatively easy to deliver, and if you can reach the people you can break through the special interests that so control America's destiny at every level. Whether it is the school, the community, or the nation, there is nothing more powerful than an idea whose time has come. And when it comes to education, cooperative learning is one of those ideas.

As with other parts of the curriculum, cooperative learning must be fashioned with a sensitivity to how students connect content areas and relate scientific, technological, and linguistic ideas with one another. With students, this means working in pairs or mixed-ability small groups to help each other learn. Students learn to take more responsibility for their own learning. There is a growing consensus that active team learning can enrich the science and language-arts curriculums, as well as helping to foster cooperative relationships among teachers, administrators, and parents. More than other innovations, cooperative learning can change school climate by encouraging cooperation, cohesion, and teamwork. Teachers who are new to the technique can learn a great deal from peers who have preceded them.

When it comes to grand curriculum designs and setting standards, it means that groups that are not traditionally thought of as within curriculum boundaries must collaborate in the construction of guidelines. "Scientific literacy" is just one example. As the definition evolves, this literacy is being viewed as the ability of citizens to come together and make decisions about social and scientific issues knowledgeably and with sensitivity to contextual values.

Increasingly, group work and problem solving (in science and language arts) revolves around literature, poetry, journal writing, art projects, drama, and song composition. Within new interdisciplinary learning environments, collaboration can help produce instruction that enhances literacy across many domains, is relevant to students' anticipated experiences, and is inclusive. In the collaborative classroom, students show more willingness to interact and reward others, there is more lasting crosscultural friendship, and there is more acceptance of students with disabilities.

ORGANIZING CONFIDENT STUDENT WORK TEAMS

Getting started with cooperative learning in language arts and science means defining student and teacher responsibilities. One of the fist steps is usually preparing students to work together. Another beginning concern is organizing the classroom so that it is easier for students to develop and practice group-process skills. Cooperative learning will not take place with students sitting in rows facing the teacher. Desks must be pushed together in small groups or replaced with small tables to facilitate group interaction. Resource and hands-on materials must be made readily accessible. Collaboration will not occur in a classroom which requires students to raise their hands to talk or move out of their desks. Responsible behavior needs to be developed and encouraged. Authoritarian approaches to discipline will not work if we are expecting students to be responsible for their own learning and behavior.

Changing the classroom organization so that students are eye to eye and knee to knee probably requires changes in the physical structure. This may mean adding worktables or pushing chairs together to form comfortable work spaces that are conducive to open communication. Other changes involve the noise level in the room. Sharing and working together even in controlled environments will be louder than an environment where students work silently from textbooks. Teachers need to learn to evaluate whether noise is constructive. Perhaps we need to learn to justify silence, not communication.

When the initial activity or problem solving is over, students need to spend time reflecting on the group work. A basic question at the end might be this: What worked well and how might the process be improved? Evaluating cooperative learning necessitates a variety of procedures. Learning outcomes will undoubtedly continue to be measured by such instruments as standardized tests, quizzes, and written exams. In addition, cooperative learning demands subjective measures. Students and teachers need to be involved in evaluating learning products and the learning climate. The learning climate involves such things as self-esteem, individual and group achievement, discipline, cooperation, values, expression, and learning together. In organizing two, three, four, or five-member groups it is best to try and organize as heterogeneously as possible. We suggest not going beyond five students in a group. Even a small group can mix children by sex, race, ethnic background, and academic ability.

Peer tutoring is a good way to introduce new vocabulary and new concepts in science and language arts. Such group settings improve student concept acquisition for everyone. Many culturally diverse learners show extra gains in cognitive growth, self-esteem, and a changed attitude toward learning (Cohen, Lotan, & Catanzarite, 1990). Social interactions are fundamental to negotiating meaning and building a personal rendition of knowledge. Effective interpersonal skills are not just important for school activities, they benefit students in later educational pursuits and when they enter the workforce.

Mixed-ability learning groups have proven effective in teaching science and the language arts. It is important to involve students in establishing rules for active group work. Rules should be kept simple and might include the following:

- Everyone is responsible for his or her own work.
- Productive talk is desired.
- Each person is responsible for his or her own behavior.
- Try to learn from others within your small group.
- Everyone must be willing to help anyone who asks.
- Ask the teacher for help if no one in the group can answer the question.

Culturally diverse learners seem to find cooperative learning activities more satisfying and useful than individual learning activities. They report a preference for groups or extended-family settings as primary learning environments (Fields, 1988). Fields also suggests that cooperative learning fosters social skills, positive peer relationships, and higher levels of self-esteem in all students. As learning teams discuss science and language arts, children can construct meaning by jointly working on solutions to problems, raising original problems, and exchanging ideas. In the process, science, language arts, and technology applications can come alive as student teams connect the concepts being studied to their everyday experiences.

Group roles and individual responsibilities also need to be clearly defined and arranged so that each group member's contribution is unique and essential. If the learning activities require materials, students may be required to take responsibility for assembling and storing them. The operative pronoun is "we" instead of "me."

THE ROLE OF THE TEACHER

In a cooperative-learning classroom, teachers need to provide time for students to grapple with problems, try out strategies, discuss, experiment, explore, and evaluate. Talking about key concepts in science and language arts is as new to some teachers as participating in small groups. The new science and language-arts standards (with their accompanying reports) encourage a break with tradition by encouraging teachers to put students in small groups where they can argue about key concepts and work on problems as a team (NSTA, 1996). The primary focus is on the students' own investigations, discussions, and group projects. The teacher's role shifts to one analogous to that of a team coach or a manager with expertise in the area. Whatever model a teacher chooses for cooperative learning, there are certain shared characteristics. Students are given opportunities to integrate their learning through group discussion, discovery experiences, practice, and other group activities. In addition, any approach to cooperative learning usually incorporates group goals, individual accountability, and an equal opportunity for all group members to achieve success.

Teachers need to model attitudes and present themselves as problem solvers and models of inquiry. They do this by letting students know that everyone is an active learner and no one knows all the answers. Teachers also need to exhibit an interest in finding solutions to problems, show confidence in trying various strategies, and risk being wrong. It is more important to emphasize the aspect of working on the problem than getting "the answer." Near the end of an investigation the teacher can develop more class unity by pointing out how each small-group research effort contributes to the class goal of understanding and solving problems or exploring a topic. By creating a safe environment where students are encouraged and affirmed, teachers can push the boundaries of the curriculum into new spaces.

A Group Investigation Model

Planning

The teacher puts forward the problem to investigate.
The teacher organizes the class for group investigation.

Implementation

The teacher prepares students by explaining cooperative learning and putting some charts up in the room that outline the process and explain roles like reader, checker, encourager, and recorder.

The concern, question, or problem to be investigated is presented.

The class brainstorms and determines topics to be studied.

The class organizes into interest-based research teams.

Individuals within the teams assume their role and the group cooperatively plans their investigation.

Research teams carry out their inquiry and organize their findings.

The group project is briefly presented to the whole class and all the groups pool information to address the original question or problem.

Follow-up

The students are held individually accountable for their work by test, portfolio, or performance assessment.

Teams are given group rewards based on individual accomplishment and shown how each group's work contributes to the class understanding of a topic.

Students reflect on their work as a group and make suggestions for improvements as they consider new investigations.

TEACHING THE COOPERATIVE GROUP LESSON

When teaching a lesson, it is important that students understand the problem situation and work together to find and evaluate the solution. Specific strategies are helpful in each of these steps.

Introducing the Lesson

During the initial introduction, the teacher's objective is for the students to understand the problem or skill and establish guidelines for the groups' work. The teacher presents or reviews the necessary concepts or skills with the whole class and poses a part of the problem or an example of a problem for the class to try. Opportunities for discussion are provided. The actual group problem is then presented after the conceptual overview. The class is encouraged to discuss and clarify the problem task. Before breaking the class into groups, it is also helpful to have a student or two explain the problem back to the class in their own words. Teachers make sure that the groups are together, help explain the task, clarify, define, and promote thinking skills.

Group Exploration

Students work cooperatively to solve the problem. The teacher observes, listens to the groups' ideas, and offers assistance as needed. The teacher is also responsible for providing extension activities when a group is done early. If a group is having difficulties, the teacher helps them discover what they know so far, poses a simple example, or perhaps points out a misconception or erroneous idea of the group. Sometimes a group has trouble getting along or focusing on what they are supposed to be doing. At this time it may be necessary to refocus the group's attention by asking questions, such as the following: What are you supposed to be doing? What is the task? How will you get organized? What materials do you need? Who will do what?

Summarizing

After the problem task and group exploration are completed, students again meet as a whole class to summarize and present their findings. Groups present their solutions and share their processes. When the processes are shared, both group procedures and problem-solving strategies are summarized.

Questions might include the following: How did you organize the task? What problems did you have? What method did you use? Was your group method effective? Did anyone have a different method or strategy for solving the problem? Do you think your solution makes sense? Encourage students to generalize from their results: What other problem does this remind you of? What other follow-up experiment could you try based on your findings? Students are encouraged to listen to and respond to other students comments. It may also be helpful to make notes of the responses on the chalkboard to help summarize the class data at the close of the lesson.

COOPERATIVE INQUIRY ACTIVITIES

The following sample activities should be done with students working in small groups. This science and math lesson is designed to help students discover the properties of water by using hands-on concrete experiments. The second set of activities spells out the materials needed and then looks at five different experiments. The procedures of the experiments were created by preservice teachers intended for upper elementary children (Nina McCauley, Jeanne Edwards, Kerry Forni, Courtney Collins, and Kathy Green at SFSU). Facts about the experiments and background information follow the activities.

Activity 1: Wind, Air, and Water
(Earth Science, Data Collection)

Materials Needed

Glass bottle, tumbler, chalkboard, piece of cardboard, shiny can, ice, hot water, alcohol, small lamp, two sponges.

Experiments

A. *Investigate the effect of wind on evaporation*
   1. Get two sponges the same size and wet them.
   2. Make two spots of equal wetness on the chalkboard.
   3. Fan one of the spots vigorously with a piece of cardboard.

B. *Discover the cooling effect of evaporation*
   1. Have students dip their forefinger into a tumbler of water.
   2. Remind students to keep their forefinger and dry middle finger a slight distance apart.
   3. Instruct students to blow on both fingers at the same time.

C. *Compare the rate of evaporation of different liquids*
   1. Instruct students to hold out both hands palms down.
   2. Put one drop of rubbing alcohol on the back of one hand and a drop of water on the back of the other hand.

D. *Observe water vapor condense*
   1. Add water to a shiny can until it is half full.
   2. Add ice cubes and stir.
   3. If it is winter, add salt to the can.

E. *Make a fog*
   1. Fill a clean soda bottle (or any narrow-neck bottle) with very hot water, adding the water slowly to prevent the glass from cracking.
   2. Now pour out most of the water, leaving about 5 cm (2 in.) at the bottom.
   3. Put an ice cube on the mouth of the bottle and hold the bottle so that it is facing the sunlight or the light of a lamp.

Facts Explained

A. The moisture on the fanned spot will evaporate more quickly because the fanning blows away the saturated air above the spot and provides a fresh supply of unsaturated air.

B. As water evaporates from the forefinger, the heat needed for evaporation is taken from the finger. The middle finger, which serves as a control, does not become a cooler.

C. The alcohol will evaporate more quickly because its molecules are moving faster from the beginning at the same temperature. In addition, because the rubbing alcohol evaporates more quickly, it takes heat away from the hand more quickly, and the spot with the alcohol on it will feel cooler as well.

D. Soon a thin film of tiny droplets of water will form on the sides if the can as the air containing water vapor is cooled and the molecules of water evaporate more slowly and come close enough together to become water again. The thin film will gradually form large droplets. In the summer, the humidity may be so high that the water vapor will condense without ice cubes having to be added. In the winter, the humidity may be so low that salt will have to be added to the cold water and ice cubes to get the water vapor to condense.

E. A fog will form in the bottle as the warm, humid air is cooled by the ice cube and the cool air below the ice cube. The water vapor condenses in tiny droplets that float in the air.

Background Information

Evaporation

When water changes from a liquid into an invisible gas, called water vapor, this change is called evaporation.

Evaporation always takes place at the surface of the water.

Evaporation takes place because of the molecular motion within the water.

- Water is made up of tiny molecules that are constantly moving.
- Some of these molecules have more energy and move faster than others.
- The faster-moving molecules near the surface of the water leave the surface to go off into the air, becoming molecules of water vapor.

Some solids, like moth balls, can evaporate directly as a solid without first becoming a liquid.

Factors Affecting Evaporation

Heat will make water evaporate more quickly.

- Heat makes the molecules move faster.
- As a result, more molecules can leave the water at one time.
- The larger the surface, the more quickly evaporation will take place because more molecules can leave the water at one time.

The amount of water already in the air affects the speed of evaporation.

- If the air already contains a lot of water vapor, there is less room in the air for more molecules of water vapor to enter, and the speed of evaporation is slow.
- If the air contains only a little water vapor, there is plenty of room in the air for more molecules to enter, and evaporation takes place more quickly.

Wind helps water evaporate more quickly.

- As the molecules leave the water and become water vapor, the air above the water eventually becomes filled, or saturated, with water vapor.
- This saturation slows down evaporation because there is no more room in the air for more molecules of water vapor to enter.
- Wind blows away the air that is saturated with water vapor and provides fresh air that can hold a fresh supply of water vapor.

The lower the air pressure above the surface of the water, the faster evaporation takes place.

- Lower air pressure means that the air is not pressing down as hard on the surface of the water.
- This lower air pressure makes it easier for the molecules to leave the water and go into the air as vapor.

Warm, dry air can hold more water vapor than cold, moist air; consequently, the warmer and drier the air above the water, the faster the water evaporates.

Liquids other than water also evaporate, and some liquids evaporate faster than others because their molecules are moving faster, from the beginning, at the same temperature.

Follow-Up Interdisciplinary Activities

1. Write a poem or short story about water in the air.
2. Record the daily temperature and take barometer readings daily.
3. Clip out the weather report and have students pretend to be a weather forecaster.
4. Have students keep a log and record differences in the weather reports as they occur.

Activity 2: Seeds Hunt
(Life Science, Language Skills)

Materials

Science and language-arts journal, brown paper bag, pencil.

Objectives

1. Students will classify seeds and the properties of seeds.
2. Students will describe the categories verbally and in writing and the reasons for putting the seeds in it.
3. Students will compare and order their seeds.
4. Students will work cooperatively together.
5. Students will make predictions (guesses) where the seed might be categorized.

Procedure

1. Introduce the concept of the great variety in plant seeds. Tell students that they are going to go on a seeds hunt with a small group of

two or three other classmates Their task is to try to find and collect samples of one seed that fits each of these categories:
- Seeds that float
- Seeds that blow
- Seeds that hitchhike
- Helicopter seeds (seeds that twirl)
- Seeds that are cycled through animals or need animals to grow
- Other kinds of seeds

2. Instruct students to record in their science and language-arts journal where each seed was located. Students may wish to describe, draw, or write their feelings about finding the seeds.
3. Upon returning, have students guess where the seed may be categorized. Again, have them write their guess in their journals.

Evaluation
1. Have student groups bring their seeds back to the class to compare their findings and test their guesses with other groups.
2. Have students divide and save the seeds in their portfolios.

COLLABORATIVE ART, SCIENCE, AND LANGUAGE-ARTS CONNECTIONS

By picking out some visually confronting pictures that suggest a connection to science or math (in magazines like *Art in America* and *Art Forum*), teachers can stimulate student writing. If you want to get closer to colorful science concepts the weekly newsmagazine of science, *Science News*, works quite well. We suggest connecting to the art world as a unifying concept for subject-matter integration (*Art News* or *Art in America*). You can cut out different pictures of art, enlarge them, make multiple copies with a color copier, and laminate. Another possibility is working from a few thought-provoking slides. Contemporary art that has something identifiable happening is best. Art, like science and language, is a gesture of the imagination.

Questions to Stimulate Discussion
(in Groups of Two or Three)

What does it remind you of?

What does the motion feel like?

What is in the shadows or behind an object?

What sounds do you hear; what colors do you see?

Imagine yourself in the painting; What do you feel?

You have entered the painting; something has joined you: What is it?
Meet people you want to meet; Who are they and what are they up to?
Does the title of the painting suggest something to you?
Make a list of at least ten things you think you see in the painting: Create something out of these observations.

Language development in children moves from the social to the personal and back to the social. In the social setting of a writing group, students can collaborate with peers and work with others to develop their writing skills by spotting problems and discussing, defining, rejecting, and accepting suggestions. The goal is to help students develop productive relations with others based on respect, trust, caring, and cooperation. Whether it is two students writing together or a small group providing feedback to a writer, the collaborative effort can be invigorating.

Roles may vary depending on the activity and number in each group. In scientific experiments, it is useful to have someone responsible for getting the materials and cleaning up. A typical group of four includes a reader, who reads the problem to the group; a checker, who makes sure everyone knows what to do; an encourager, who keeps it on task and lively; and a recorder, who explains what went on to the whole class. If there are three in the group, then everyone shares the encourager role. With two, each get two roles. An example of a group worksheet to facilitate this process is shown on page 46. For students to successfully complete their cooperative learning tasks, they need to negotiate, compromise, cooperate, and arrive at a synthesis based on rational thought.

COOPERATIVE STRUCTURES

Cooperative learning can help us with social values and subject-matter accomplishment. Research has clearly shown that cooperative learning leads to significant student gains on measures of academic achievement, self-esteem, social development, and crosscultural relationships (Slavin, 1989; Sharan & Shachar, 1988).

The ways we choose to spend our teaching and learning time conveys implicit messages to students about what is valued and important. The bulk of time currently spent in classrooms is spent listening to the teacher or working on isolated paper-and-pencil tasks usually found in textbooks or on practice sheets. If large amounts of time are devoted to the learning of isolated skills, the underlying assumption conveyed is that learning means mastering a narrow range of skills and emphasizing product. Important educational matters, like literacy,

Group Worksheet

Names of group members: _____

_____

Group role of each individual:

| Checker | Encourager | Recorder | Materials | Clean up |

Assignment or group goal: _____

_____

_____

List what has to be done to complete the goal:

1. _____

2. _____

3. _____

4. _____

5. _____

List the materials needed to complete this assignment:

1. _____

2. _____

3. _____

Due dates: _____

---

are much broader, and more an attitude toward the world than a set of subskills to be mastered.

Competition, anxiety, a product orientation, and lack of emphasis on problem solving have been identified as problems that inhibit children's development. Schools have traditionally been competitive institutions, with increased emphasis on competition as students move up through the system. The messages sent through this model is that quick, "right" answers usually drawn from short-term memory are valued. Children learn to depend on the "faster" students to provide answers and on the teacher for validation of their thinking. Focusing on completing assignments quickly, children rarely stop to question the reasonableness of their response or the meaning behind it. Such pedagogical techniques not only inhibit students' self-reliance, but do little to produce an understanding of science and mathematics.

Research suggests that high levels of anxiety significantly inhibit achievement (Jones & Johnson, 1990) (a little anxiety may actually stimulate interest). When a competitive structure is bonded to an abstract style of presentation, anxiety is a major product. In this environment, many students withdraw into themselves as they mentally look for desert islands that are unreal but there. Cooperative learning attempts to deal with these concerns by maximizing interaction and changing what is valued in the learning process. The result can be classrooms where learning is both more possible and more productive.

Working in collaborative groups gives students a better chance to explore ideas, justify their views, and synthesize knowledge within a supportive environment. Instead of quiet, isolated "workers" reluctant to share answers and shielding their papers from other students' eyes, students are encouraged to share ideas, collaborate together, and pool their knowledge to solve or perhaps resolve a problem. Learning is more cooperative and less competitive.

Professional associations have long called for more of an emphasis on group learning and relying on reason more than rote. The American Association for the Advancement of Science (AAAS) (1990), for example, has called for a collaborative common core of learning that stresses thinking skills over memorization of details. One of the key features of cooperative learning is that students pool critical thinking, giving and receiving help. Webb (1982) notes that such interaction correlates with critical thinking and risk taking. Teaching young people to think and work in groups are educational goals expressed repeatedly by teachers and in the literature. Yet they remain largely unmet in some classrooms, despite the evidence that small-group collaboration can lead to superior achievement in higher-order thinking (Hertz-Lazarowtiz, Sharan, & Steinberg, 1980; Sharan & Shachar, 1988).

In developing collaborative learning structures, it is necessary to create a learning environment in which it is safe to make mistakes and learn from those mistakes. Another structural element in cooperative learning is maximizing social interaction and focusing on learning that is exciting and pleasurable. Such environments do not require adapting one's thinking to the person "in charge," but encourage contributing to a shared understanding of something and finding ways to share that knowledge.

Once the curriculum of the past is viewed from a safe distance (like Dick and Jane in the old reading textbooks), it can be seen as a quaint anachronism. But while your professional life is squeezed into the narrow confines of yesteryear, the charm escapes you. Traditional goal structures tended to be teacher centered. Teachers controlled learning by imparting knowledge, maintaining control, and validating thinking. They controlled the learning that went on, knowing the end product of the learning experiences as well as the sequence of activities and procedures to achieve the learning goal.

Successful implementation of cooperative learning requires deep conceptual understanding of the process by both teachers and students. For students to actively collaborate in small groups involves significant changes for many classrooms. The teacher is faced with the difficult task of encouraging students to become responsible for their own learning. One of the goals is to have students rely more heavily upon their classmates for assistance in doing a task and evaluating an answer. There is no waiting around with hands up; students get immediate help from their peers. Teachers specify the instructional objectives, arrange the classroom to maximize social interactive, provide the appropriate materials, explain the task and the cooperative goal structure, observe the student interactions, and help students solve problems. They pay attention to the learning process and social relationships within the groups. And they evaluate the group products.

In a collaborative setting the teacher helps children gain confidence in their own ability and the group's ability to work through problems and consequently rely less on the teacher as the sole knowledge source. Students are motivated more by social contact with their peers and by their sense of achievement as they succeed in challenging tasks through group effort rather than through strict, step-by-step teacher direction. Technology can amplify the process. Science has produced new communications technology (like the Internet) that allows for collaboration on a global scale. We now have both the opportunity and the power to shape the new rules we want to live by.

Students in cooperative learning settings often raise questions and ideas which go beyond the teacher's guide. To keep up to date, teachers must also continue being learners and become comfortable with saying "I don't know" or "let's find out" as students push them in new learning directions unimagined and unplanned. Cooperative classroom environments require teachers who actively seek to create them. When they do, many teachers will find that their best instincts about group work are confirmed.

CHANGING STUDENT ATTITUDES

Students must also undergo a major shift in values and attitudes if a collaborative learning environment is to succeed. Getting over years of learned helplessness will take time. The traditional school experience has taught many students that the teacher is there to validate their thinking and direct learning. Since entering school, children are compared with one another for recognition, and even the smallest detail is often decided by the teacher. Unlearning these dated modeling structures takes time. Students used to traditional classrooms will need time to adjust to group work and decision making.

It is important that students understand that simply "telling an answer" or "doing someone's work" is not helping a classmate learn. Helping involves learning to ask the right question to help someone grasp the meaning, or explaining with an example. These understandings need to be actively and clearly explained, demonstrated, and developed by the teacher.

Attitudes change as students learn to work cooperatively rather than taking individual ownership of ideas. As they share rather than compete for recognition, many children find time for reflection and assessment, as opposed to rushing to finish a task. Small groups can write collective stories, edit each other's writing, solve mathematics problems, correct homework, prepare for tests, investigate science questions, examine artifacts, work on a computer simulation, brainstorm an invention, create a sculpture, or arrange a new rap music tune. Working together is also a good way for students to synthesize what they have learned, collaboratively present to a small group, coauthor a written summary, or communicate through the subtleties of the arts. It is little wonder that operative learning often becomes more a "culture" or set of collaborative values (pervading the classroom) than a technique.

BENEFITS FOR STUDENTS AND TEACHERS

A major benefit of cooperative learning is that students are provided with group stimulation and support. The small group provides safe opportunities for trial and error as well as a safe environment for asking questions or expressing opinions. More students get chances to respond, raise ideas, or ask questions. As each student brings unique strengths and experiences to the group and contributes to the group process, respect for individual differences is enhanced.

The group also acts as a motivator. Many times ideas are pushed beyond what an individual would attempt or suggest. The quality and quantity of thinking increases as more ideas are added, surpassing what the individual could do alone. Group interaction enhances idea development and students have many opportunities to be teachers as well as learners. Simultaneously, the small-group structure extends children's resources as they are encouraged to pool strategies and share information. More withdrawn students become more active. Students who often have a hard time sticking to a task receive group assistance so they can learn to monitor their time better and become a productive member of the group. The unity of the group has been found to extend beyond the classroom, to the playground and social situations (Adams & Hamm, 1989).

Students monitor each other while creating a spirit of cooperation and helpfulness. Students must agree to collectively ask for help, so they ask better questions and are more eager for teacher input. Cooperative learning can help teachers spend less time being police, as students learn they are capable of validating their own values and ideas. Teachers are freer to move about, work with small groups, and interact in a more personal manner with students. Cooperative group learning can also be arranged so there is less paperwork for the teacher. Six or eight group papers is usually less of a problem than twenty-five or thirty. In this structure, teachers continue to be learners, opening channels that may never have been imagined.

CHANGING TO A COOPERATIVE LEARNING MODEL

It usually takes more than one or two tries with cooperative learning to get cooperative groups going. Those not used to active-learning teams must be eased into the process through a consistent routine. When high achievers from traditional classrooms have to work in a group they may have initial problems because they are accustomed to being rewarded for quick answers with low levels of thinking. It may take some time for them to become comfortable working cooperatively. Teachers and administrators have had to make a conscious effort to move schools away from a competitive factory model. This meant going beyond the usual cosmetic effects (that have little positive value) to significant structural changes.

Having good models can help, but major change will require an emphasis on visions and a sizable national experiment. It takes time and practice for the vital energy inherent in new skills to become part of a teacher's repertoire. Teachers need to be actively involved in learning new strategies and goal structures, and they need support while implementing new skills. Change takes time and is a result of systematic staff development. Inservice workshops can help provide assistance as teachers try activities, share experiences, and receive feedback within a supportive environment of collaboration. Making positive changes in the organization of learning requires an environment where it is safe to make mistakes and learn from those mistakes.

Cooperative learning has a large and supportive empirical base at the elementary level, where teachers seem to take naturally to it. The research is weakest at the high school and college levels, where attention is now being given to how collaboration can lead to higher-level conceptual learning. When it comes to inquiry and problem solving in science and language arts, cooperative learning has proven itself as effective. In the elementary school, cooperative learning may very well be the single most effective way to bring about change in a traditional school environment. However, the results are not automatic; it takes sustained professional development activities and inservice training for teachers to rearrange their classrooms to stimulate comprehension and higher-order thinking.

As students and teachers learn to conduct their communications with a collegiality (bordering on civility), collaborative teams can help build a broader sense of community and even stimulate collaboration between teachers. Connecting mutual achievement and collegial relationships may be extended to staff meetings, committees, and relationships with parents. Traditional teachers and parents must come to understand that it is sometimes all right for students to share answers. As students and teachers become participating members of a collegial team of peers, both develop a joint sense of purpose and even more efficiently tap the possibilities of cooperative learning.

Opponents will have to be satisfied with attacking the tendency to oversell and undertrain, and the occasional tendency to focus more on social settings than what is taught. Method certainly is not all, but in general the empirical and practical evidence supports the use of cooperative learning techniques in the classroom. And once the main elements are understood, teachers can adapt cooperative learning to the needs of their students and their unique classroom situation. It is now generally accepted as a useful tool for helping students get the most out of exploring, experiencing, problem solving, and finding meaning in science and the language arts.

To gain and share expertise, members of a team challenge each other's thinking in a way that does not breed conformity. Work teams can be used for the majority of instructional time at any grade level to provide students with a support network. In cooperative learning communities, instruction can become personalized and mutual achievement and caring for one another really are important. Students and teachers can come to view one another as collaborators who help one another cognitively, emotionally, physically, and socially. Active team learning is more than an innovation in itself, it is a catalyst for other changes in curriculum, instruction, and schooling. Slavin (1990) goes so far as to suggest that becoming a contributing member of a collegial team promotes self-discovery and higher-level reasoning.

In cooperative learning communities, the teachers might be thought of as community leaders and the students might be viewed as citizens. Students and teachers can be rooted in a network of familial and community relationships that make up a civil society. Like an extended family, everyone cares about individual and mutual achievement. Individual rights are balanced by reciprocal obligation and mutual interdependence.

By tapping into a student's natural curiosity and creating a learning community, students and teachers can use cooperative methods to achieve academic goals. Cooperative learning can be used with confidence across subjects and grade levels to explore meaning and help students care for one another. As students come together in teams, they can use communication skills to look at scientific issues and jointly ask, "Where can we go with this?" and "How might it make our world better?" By building on the group energy and the idealism of youngsters, the thinking, learning, and doing process can be pushed forward.

What children can do together today, they can do alone tomorrow.
—Vygotsky (1962)

Chapter 4

Assessing Student Understanding

You don't get motivation by applying rewards and punishments. You get motivation by bringing people into the process and helping them feel as though they're on internal control.
—Stiggins (1997)

Assessment and evaluation are so intertwined it is hard to separate them. Assessment is collecting data to make a judgment. Evaluation is judging the value of something based on the available data. You can have assessment without evaluation, but you cannot have evaluation without assessment. The concern here centers on the role of performance assessment in the evolving definition of literacy. The focus is on using portfolios in conjunction with interdisciplinary inquiry.

There is general agreement that the methods for assessing educational growth have not kept up with the new subject-matter standards and the way the curriculum has changed. Multiple-choice testing just does not do a very good job of capturing the reality of today's students. Such tests convey the idea to students that bits and pieces of information count more than deep knowledge. On the other hand, assessing performance conveys the notion that reasoning, in-depth understanding, taking responsibility, and the ability to apply knowledge in new situations is what counts.

Assessment can be used to motivate students in several ways. To begin with, children can be assigned to cooperative teams for interdisciplinary inquiry and peer assessment. As students are brought into designing assessment procedures as responsible partners, the whole process is enhanced. Students can then use portfolios to keep their own records and reflect on how well they are doing. By viewing the evidence of their increasing proficiency, they become reflectors of their own progress. Finally, learning to communicate with peers, teachers, and parents about their achievement means taking more responsibility for academic success.

USING GROWTH PORTFOLIOS

Many teachers share a vision of what they think should be happening in their classes. It goes something like this: Students work in small groups doing investigations or accomplishing tasks using tools such as manipulative materials, blocks, beakers, clay, rulers, chemicals, musical instruments, calculators, assorted textbooks, computers, the Internet, and other references. They consult with each other and with the teacher, keeping journals and other written reports of their work. Occasionally, the entire class gathers for a discussion or for a presentation. They want students to be motivated and responsible. Traditional testing methods do not support this vision.

As students and teachers use product criteria (the performance or work samples) and progress criteria (effort or class participation), they conduct experiments, collaborate in interdisciplinary projects, and construct portfolios. Portfolios represent a more authentic and meaningful assessment process. They are a major performance-assessment tool for having students select, collect, reflect, and communicate what they are doing. Having children think about the evidence they have collected—and decide what it means—is clearly a good way to increase student engagement.

Portfolios have long been associated with artists and photographers as a means of displaying collected samples of representative work. They have also been used for over ten years by various reading and writing projects (Graves, 1994). In the 1990s, the National Assessment of Educational Progress (1995) has suggested using portfolios to assess students' writing and reading abilities. They usually begin by specifing the essential concepts to be covered, figuring out how to link what is taught to the assessment process, and finding ways to display an understanding of the results. What is new is that the interest in using these performance-assessment techniques now stretches across the curriculum.

Teachers have found that collecting, organizing, and reflecting on work samples ties in nicely with active interdisciplinary inquiry. Portfolios not only capture a more authentic portrait of a child's thinking, but can serve as an excellent conferencing tool for meetings with children, parents, and supervisors. In addition to portfolios, teachers often create other performance-assessment tasks: projects, exhibitions, performances, and experiments. By creating opportunities for students to reveal their growth, we help them understand what they are doing and why they doing it.

DRAWING MEANING FROM WHAT IS OBSERVED OR MEASURED

Assessment is a broader task than testing because it involves collecting a wider range of information that must be put together to draw meaning from what was observed or measured. Of course, the first use of assessment is within the classroom, to provide information to the teacher for making instructional decisions. Teachers have always depended on their own observations and examination of student work to help in curriculum design and decision making. Teachers need ongoing support in their efforts to set high goals for student achievement.

Lately we have been hearing a lot about "authentic assessment." The term implies evaluating by asking for evidence of the behaviors you want to produce. For assessment to be authentic, the form and the criteria for success must be public knowledge. Students need to know what is expected and on what criteria their product will be evaluated. Success should be evaluated in ways that make sense to them. It allows students to show off what they do well. Authentic assessment should search out students' strengths and encourage integration of knowledge and skills learned from many different sources. It encourages pride and may include self and peer evaluation.

In the world outside of school, people are valued for the tasks or projects they do, their ability to work with others, and their responses to difficult problems or situations. To prepare students for future success, both curriculum and assessment must encourage this kind of performance assessment (Stenmark, 1989). Assessment of products that students produce may include portfolios, writing, group investigations, projects, interactive websites, class presentations, or verbal responses to open-ended questions. Whether it is small-group class presentations, journal writing, storytelling, simple observation, or portfolios, alternative assessment procedures pick up many things that children fail to show on pencil-and-paper tests.

PURPOSEFUL ASSESSMENT

Learning demands communication with self, peers, and knowledgeable authorities. It also asks for effort and meaningful assessment. A lesson from 20th-century physics is that the world cannot ultimately be objectified. The same is true of people. Still, whether in science or education, we need to know where we are going, how we are going to get there, and how far we have progressed. Though goals and accompanying testing may be limiting, school reform requires identifying and measuring some of the desired results at various levels of schooling. Comparisons sometimes prod schools to improve. Like movie reviews, school rankings can make or break the reputation of a school. Public shaming may be embarrassing and painful, but it often leads to improvement. National tests, goals, and subject-matter standards help us know where we are going. The next step is to be sure that highly competent teachers are given the chance to get us there.

Standardized multiple-choice tests have not shown themselves to be all that helpful for teachers facing complex and multifaced socioeducational problems. Portfolio assessment is more dynamic and does a better job of demonstrating proficiency or progress toward a preset purpose. It can be used to meet the needs of evolving definitons of literacy, allowing students and teachers to create, reflect, evaluate, and act upon material that is highly thought of by those most directly involved. This helps us all go beyond simply recognizing that a mistake was made to imagining why, getting feedback from others, and finding practical ways to do something about it.

THE PORTFOLIO AS A TOOL FOR UNDERSTANDING

A portfolio is best described as a container of evidence of someone's skills and dispositions. It is a portfolio, not an archive, so avoid being trapped under an avalanche of information. More than a folder of a student's work, portfolios represent a deliberate, specific collection of an individual's important experiences and accomplishments. The items are carefully selected by the student and the teacher to represent a cross-section of a student's creative efforts. It is not just the best stuff, it is what is most important to all concerned. Portfolios can be used as a tool in the classroom to bring students together, to discuss ideas, and provide evidence of understanding. The information accumulated also assists the teacher in diagnosing learners' strengths and weaknesses. It is clearly a powerful tool for gaining a better understanding of student achievement, knowledge, and attitudes.

The portfolio-assessment process helps children become aware of their learning history and the development of their reasoning ability. Prospective teachers can find out a great deal about themselves by sketching the most important events and efforts in their school days. As students become directly involved in assessing progress toward learning goals, the barrier between the learner and the assessment of the learner is lowered. Through critical analysis of their own work—and the work of peers—students gain insight into many ways of thinking about and resolving problems.

Portfolios are being used by teachers to document students' development and focus on their growth over time. The emphasis is on performance and application, rather than on knowledge for knowledge's sake. Portfolios can assist teachers in diagnosing and understanding student learning difficulties. This includes growth in attitudes, thinking, expression, and the ability to collaborate with others. There is clearly more to learning than multiple choice. Assessing the student over time (with portfolios) brings academic progress into sharp focus and promotes reflection on the larger issues of teaching and learning.

THE DESIGN AND PURPOSE OF A PORTFOLIO

Portfolios are a means of bringing together representative material over time. Building such a picture of students' understanding is not a one-time collection of examples. This material should reflect students' knowledge and performance. The purpose is to gain an accurate understanding of students' work, development, and growth. The intent is to recapture the past to more effectively shape the future. Careful attention should be given to the following:

- What is being evaluated.
- The purpose of the portfolio.
- The appropriateness of the contents to what is being assessed.
- The audience for which it is intended.

The purpose behind the portfolio should determine its design. The range and the depth of the portfolio can be determined by the teacher, the student, or the nature of the contents. Students may choose the contents based on specific categories. For example, teachers could ask students to select examples of their work that fit into categories, such as the following:

- A sample that reflects a problem that was difficult for you.
- Work that shows where you started to figure it out.
- A sample that shows you reached a solution.
- A sample that shows you learned something new.
- A sample of your work where you need to keep searching for ideas.
- Two items you are proud of.
- One example of a comical disaster.

Involving students in the selection process gets them directly participating in their own learning and their own evaluation. Learners, teachers, and parents can gain a better understanding of the student in and out of school because portfolio contents can reveal a surprising depth of thinking and provide insights into personal issues. Collecting, selecting, and reflecting on their school experience (and that of their peers) allows students to communicate who they are and how they view themselves in relation to others.

The following items might be put in portfolios:

- Group assignments and ideas.
- Teacher assessments and comments.
- Student writings and experimental designs.
- Student reflections, journal entries, reactions, and feelings.
- Research, collected data entries, and logs.
- Problems and investigations.
- Individual and group projects.
- Creative expressions (art, photographs, audio and videotapes).
- Rough drafts and finished products.

Items selected should be dated and described. A cover letter by the author, a table of contents, and a description of the assignment or task are other ways of assisting the reader. As the portfolio develops, the contents can be added to, deleted, improved, revised, edited, or discarded.

The following are some portfolio tips:

- Portfolio assessment is not just for younger students. Portfolios can be used with students from kindergarten through graduate school.
- Assessment should be integrative and oriented toward critical thinking and solving problems, not simply recall based. These may include observational notes by the teacher and student self-assessment.
- Portfolios should show evidence of creative and critical and thinking.
- Portfolios should show the quality of activities and investigations.
- Portfolios should contain a variety of approaches and investigations.
- Portfolios should demonstrate understanding and skill in situations that parallel prior classroom experience.

Here are some suggestions for getting started:

- Primary teachers often use large boxes to contain student portfolios. Children can place their written work in folders by subject area and the folder is put in a decorated box with the student's name. Boxes are stacked for easy access and neatness.
- For older students, a three-ring binder is most frequently used for items such as oral-history interviews, copies of historical documents, photos of community service activities, worksheets, and class notes. Handouts can be three-hole punched and added, along with journal entries, written comments, quizzes, and other documents.

- An artist's folder is useful for gathering things like video cassettes and three-dimensional kinds of projects. Photographs can be taken of large projects and videotapes made of others.

USING PORTFOLIOS FOR PROFESSIONAL DEVELOPMENT

Student input adds to the process of professional knowledge building and assures the internal commitment of those involved. We ask our students to contribute to our development as teachers by writing down two things they like about some aspect of the class and one thing that we could do better. Only one complaint is allowed! Being curious about how students think and learn has always been part of a teacher's job description. What is new is including ourselves and systematically attacking the issue, sharing the information, and looking for more professional development opportunities.

Information and observations garnered from your own classroom ("action research") is part of performance assessment and can be used to improve practice. This approach can help the best teacher in the country make better informed choices about learning activities and teaching methods. As far as student research is concerned, all projects require the formulation of a good problem (one that can be stated in question form). They also require focusing on the relationship between variables, and some way of objectively testing the hypothesis. The factors (variables), like time or group, should be kept constant. Some teachers find it best to work with another teacher to collect the data, organize them, and figure out a way to share with others. This might mean graphing results, mapping out a teaching strategy, video sharing of classroom events, informal readings of class essays, or simply a few overheads to share at a faculty meeting.

The construction of teaching portfolios has to be a teacher-designed process to get a real conversation about teaching going. We need to be constantly reminded that reimaging teaching is a continuing process that looms larger than the success or failure of any single lesson or project. It connects to teachers taking charge of their professional growth and development. Taking risks is essential for change and must be considered an asset—even when everyone agrees there has been a miscue. A few of the entries in a teaching portfolio might well be lessons that went astray yet still revealed important insights into future possibilities. Far from using this to criticize a teacher, administrators are professionally required—like doctors, lawyers, or clergy—not to use this information in a harmful way. If it is used to criticize, you can be sure that the next year's teaching portfolio will be designed to cover up weaknesses rather than an instrument for positive self-initiated change.

If the highest aim of a captain was to preserve his ship then he would stay in port.
—Thomas Aquinas (trans. 1952)

The level of courage needed for the self-revelation for a good teaching portfolio requires a supportive, caring environment where successes are celebrated and failures are seen as possibilities for growth. The objective of a teaching portfolio must be clear to everybody and it is simply not ethical to put something together for development and then have it misused for formal evaluation. If the goal is to assist professional development and help the teacher in self-motivated growth, then this must be adhered to.

PORTFOLIO ASSESSMENT EXAM

Portfolios can help provide an ongoing conversation about processes relating to teaching and learning. They can also help us attend to subjects that do not lend themselves to traditional testing methods (Mitchell, 1989). Portfolios also assist in exploring what is going on in and between subjects. These samples, drawn from different times and contexts, can serve as an ongoing means of getting people talking and learning across disciplines.

A PORTFOLIO EXAM FOR PROSPECTIVE TEACHERS

Consider your work as a whole and select significant pieces from your notebook, readings, class experiences, and activities assignments. Each item should encourage you to reflect on your performance and illustrate each of the following categories. Discuss why you included each sample, how your ideas have changed over the course of the semester, and how your understanding of the subjects covered has grown over time.

Category 1

How has your knowledge of teaching grown or changed? Include an example that reflects a growth in personal understanding and samples of activities that were meaningful to you (or enlarged your understanding).

Category 2

How have your attitudes, beliefs, and personal confidence about educational concepts changed? Samples might include questions you raised or beliefs that you held at the beginning of the semester and a comparison with later ideas. This category may include some reflection on your views and how these attitudes will translate into your classroom practice.

Category 3

How have your experiences—in and out of the classroom—helped you understand yourself and the learning process? What have you found out about your personal learning style, strengths, weaknesses, self-esteem, and group communication skills? What does this mean for your students in the future?

Category 4

What relationships or connections do you see in your work (all subjects) this semester? Look over all your notes, handouts, activities, readings, assignments, and so on and select pieces that make meaningful connections for you or form relationships between concepts that you have not explored before.

Category 5

How can what you have learned be applied to elementary school students with special needs? Include personal insights and samples from children.

LINKING ASSESSMENT WITH INSTRUCTION

Portfolios are proving useful in linking assessment with instruction at every level because they allow students and teachers to reflect on their movement through the curricular process (Mumme, 1990). They also provide a chance to look at what and how students are learning, while paying attention to students' ideas and thinking processes. We do not suggest that the "pure" objectivity of more traditional testing has no place in the classroom. Rather, we must respect its limits and search for more connected measures of intellectual growth. But there is no question that when coupled with other performance measures, like projects, portfolios can make an important contribution to new literacies.

Rules for Assessing and Evaluating

1. Conduct all assessment and evaluation in the context of learning teams
   - You must assess each student's achievement, but it is far more effective when it takes place in a cooperative setting.

2. Provide continual feedback and assessment
   - Learning groups need continual feedback on the level of learning of each member. This can be done through quizzes, written assignments, or oral presentations.
3. Develop a list of expected behaviors
   - Prior to the lesson
   - During the lesson
   - Following the lesson
4. Directly involve students in assessing each other's learning
   - Group members can provide immediate help to maximize all group members' learning.
5. Avoid all comparisons betwen students that is based solely on their academic ability. Such comparisons will decrease student motivation and learning.
6. Use a wide variety of assessment tools.

Cooperative Learning Activity and Assessment Grid

Risky Business

A recent survey in Dun's Review lists the most perilous products or activities in the United States (based on yearly death statistics). Your group task is to rank each item in order of dangerousness according to the number of deaths caused each year. Place number 1 next to the most dangerous, number 2 next to the next most dangerous, and so forth.

_____ mountain climbing	_____ surgery
_____ swimming	_____ smoking
_____ railroads	_____ motor vehicles
_____ police work	_____ pesticides
_____ home appliances	_____ handguns
_____ alcohol	_____ bicycles
_____ nuclear power	_____ fire fighting

Answers: 1. smoking, 2. alcohol, 3. motor vehicles, 4. handguns, 5. swimming, 6. surgery, 7. railroads, 8. bicycles, 9. home appliances, 10. fire fighting, 11. police work, 12. nuclear power, 13. mountain climbing, 14. pesticides.

Assign students to mixed-ability teams. Be sure that the team takes charge of the quality of the work of its members. Team members must

—learn how to define and organize work processes.
—assess the quality of the processes by recording the indicators of progress.
—place the measures on a quality chart for evaluating effectiveness.

An important assessment part of the activity is an observational record that is kept by one member of the group or by the teacher. It can be arranged on a grid and may be marked during the activity. (Keep it simple so that the rater can participate.)

Observation Form

Group	Explaining Concepts	Encouraging Participation	Checking Understanding	Organizing the Work
1	_____	_____	_____	_____
2	_____	_____	_____	_____

INVOLVING STUDENTS IN PERFORMANCE ASSESSMENT

With performance-based assessment, students demonstrate what they can do by performing a procedure or documenting a skill. Performance assessment also involves students in the evaluation of their own work. Educators no longer view assessment as residing only with the teacher or as an end point to learning. As teachers and students work together to design assessments, everybody develops an understanding of what needs to be learned. Performance assessment is not new or difficult. Basically, it is something that good teachers have always done: questioning and observing students to evaluate their progress and then modifying instruction based on these observations.

Through performance assessment, teachers are encouraged to do the following:

- Incorporate assessment into the teaching and learning process.
- Use their own judgment when evaluating learning.
- Establish criteria to assure reliability.
- Describe the skills, attributes, and qualities to be developed.
- Focus on important concepts and problem-solving skills, rather than memorization of facts.
- Design tasks to provide opportunities for students to perform, create, or produce a satisfying product.

- Involve students in the evaluation of their work.
- Build commitment to change.

The object of performance assessment is to look at how students are working as well as at the completed task or product. An observer or interviewer may stay with the group or make periodic visits. Activities may be videotaped, tape recorded, or recorded in writing by an outside adult, the teacher, or the students. Looking at student performance gives teachers information about students' abilities to reason and raise questions. This type of group assessment focuses on finding out what students know and building experiences around that information. This involves student groups working together gathering information and supporting their data.

Often, teachers construct a chart at the beginning of a unit, showing what students already know, what they want to find out, and how data were collected as evidence. At the beginning of each unit of study, teachers measure students' understanding and establish learning objectives by also writing them on the chart: "What our class wants to find out." Next, the teacher introduces hands-on activities. Through first-hand experience, students discuss and develop their own meanings. Throughout the unit, teachers and students refer to the chart to find out how thinking and questions have changed, and what new questions have arisen. At the end of the unit, students complete the chart, stating what they have learned as supported by the gathered data. This knowledge is then incorporated into group writings entitled, "What our group learned." Student groups meet as a class to discuss their findings, compare and review concepts learned, and check to see if they met all the objectives. This gives the teacher insights into how well students are concentrating, how they communicate, and how well they work together at organizing and presenting information.

It is as hard to measure love for logical thinking as it is to measure the creative imagination. How do you assess tapping into natural beauty and detail (in the present) and moving toward an informed vision of the future? While far from perfect, performance-assessment techniques would seem to be the best way to do this, while documenting the degree to which a child thinks of himself or herself as an observant investigator.

CHANGING SCHOOLS MEANS CHANGING ASSESSMENT

Generating the energy to change school culture and act on new practices is an art. It has a lot to do with understanding human nature, establishing a readiness for change, and developing an intellectual understanding of new practices. The most effective innovations are usually research based and classroom friendly. Change is a personal process, but things work out best when supported by a team approach. School reform without teachers is a contradiction in terms. If their voice is left out, then we will miss a real opportunity. To move educational reform from talk to action requires the involvement of informed teachers who can use new concepts.

Authentic assessment requires students to demonstrate the desired procedure or skill in a real-life context. It must also be informed by important elements of instructional practice. When it comes to the portfolio model, teaching, learning, and assessment are intimately connected. This performance-assessment technique allows students and teachers to display growing strengths, rather than simply exposing weaknesses. The portfolio approach promotes ownership of learning by encouraging students and teachers to use and shape knowledge as they see fit. It works as well for the teachers professional development as it does for students. When good new approaches like this are shared with others, everyone gains. For teachers, sharing might involve collecting and organizing data, graphing the results, presenting a teaching strategy, and an informal discussion of authentic assessment with a colleague. Remember, when you set out to improve something it is important to have a goal in mind and room for serendipity. When it comes to possibilities for practice, we have to be ready for chance finds. Who, for example, would have predicted today's Internet based on the 1980s original?

Professionals do not blossom when they spend their careers enfolded in the logic of others. Teachers themselves are in the best position to identify the key issues, question, probe, try to attain clarity, and do something in an ambiguous world. Informal belief systems are just as important as methodology. When teachers become students of their own learning, they discover the inconsistencies between what they believe about gaining knowledge and how they practice teaching. Discoveries they make themselves are more convincing and make them more willing to change what they do. The thoughtful reflection about practice that comes with performance assessment can play a major role in helping teachers become autonomous professionals.

I believe that we should get away altogether from tests and correlations among tests and look instead at more naturalistic sources of information about how peoples around the world develop skills important to their way of life.
—Howard Gardner (1993)

Chapter 5

Scientific Literacy: Involving Students in Scientific Inquiry

Science distinguishes itself from other ways of knowing and from other bodies of knowledge through the use of empirical standards, logical arguments, and skepticism.
—National Research Council (1995)

Science is not just for scientists or poetry just for poets. As teachers, it is our job to open children's minds to the wonders of the linguistic and the natural world. One of the most important goals of science instruction is to expand the perception and the appreciation of water, rocks, plants, animals, people, and other elements of the world around us. The next step is being able to use technological tools to communicate those understandings. The technological products of science are important for many reasons, not the least being their effect on human communication. Using the technological and intellectual tools of science (scientific processes) to explore that world can spark curiosity, interest, knowledge, and action.

By placing science and its literacy associates closer to the center of the school curriculum, we can help students develop a sense of wonder about the world in which they live. Observing, measuring, classifying, inferring, and communicating come naturally. As children sharpen their powers of observation, they become capable of noticing things that they had not noticed or thought of before. In school, this might mean participating in project constructions that build on the physical principles being studied. Out of school, students might act as "nature detectives" to explore a city park or wet places (like ponds), looking for plants and animals they might not have noticed before. They can record their observations like scientists. Whether inside or outside, when you teach science with active inquiry and problem-solving methods you cannot help but get closer to the new standards and goals of science education.

As you read this chapter, think about the following:

- What science is.
- Scientific literacy and its implications for the classroom.
- The science standards for elementary teaching.
- Examples of the process skills of language arts, science, and technology.
- How children learn science and how their misconceptions are formed.
- Using some of the activities and methods in this chapter to enrich your language-arts and science curriculum.

SCIENTIFIC LITERACY FOR ALL STUDENTS

Science education helps students develop understandings and ways of thinking that are essential for all citizens. We view scientific literacy as being familiar with the following:

- The natural world, recognizing its diversity and unity.
- Important ideas and rules of science.
- Some of the important ways science, communication, and technology depend on each other.
- Science and technology as human enterprises, and what this implies about their strengths and weaknesses.
- Scientific processes and ways of thinking.
- How scientific literacy might positively influence personal interactions and societal decision making (AAAS, 1990).

All students can learn science, and all students should have the chance to become scientifically literate. This is one of the themes of the National Science Education Standards (National Research Council [NRC], 1995). The idea of scientific literacy, in one form or another, has been around for the better part of this century. Spurred on by John Dewey's (1916) educational theories, science education and accompanying notions of scientific literacy gained a stature on a par with reading, writing, and arithmetic. Whether viewed from a cultural or a practical vantage point, science has long been seen as something important that has not succeeded in getting through to the majority of our citizens. It will take more than changing the schools. Science literacy must begin in the early grades, when students are naturally curious about their world and eager to explore it. Another theme in the standards is that learning is an active process achieved by enthusiastic and motivated students. As far as science educators are concerned, there are now models for effective science instruction that help teachers move students along the road to scientific literacy.

Science is usually portrayed to the public as the most rational of human enterprises. Yet the reasoned approach that supposedly characterizes the practice of science has had little effect on the American public. In *The Myth of Scientific Literacy*, Morris Shamos (1995) points out that in spite of all the effort, by any reasonable measure we remain a nation of scientific illiterates. Scientific literacy extends beyond concepts and procedures to include the history of scientific ideas and the nature of science and technology. It also means understanding how these closely related domains effect personal life and our society. Achieving anything like excellence in the multidimensions of scientific literacy requires and deserves the combined and sustained support of all Americans.

Physical, life, and earth science are good examples of conceptual and procedural knowledge that relate to scientific literacy. The new standards include the processes of science and give greater emphasis to cognitive abilities like using logic, evidence, and extending their knowledge to construct explanations of natural phenomena. The science standards encourage scientific literacy by revealing the basic concepts and abilities that all students should develop.

Scientific reasoning is most likely to become part of an approach to lifelong learning if it is frequently applied to problems encountered in life. If early work in science offers models that are found useful in future experiences, then a scientifically literate citizenry is not beyond the realm of possibility. One of the important goals in elementary science is to provide a firm foundation for linking what students are learning in science to the activities normally pursued by children and adults. Making what students learn in school relevant to their lives has all kinds of positive consequences.

CHANGING THE SCIENCE CURRICULUM

The structure of science education, as we know it today, was more or less defined in the 1960s. The 1970s added the idea of a unified science education (combining the various sciences), and the 1980s introduced educators to the STS (science/technology/society) components of the science curriculum. In the 1990s, the national standards for science education were constructed on these foundations. Early on, it was found that children learn best when they actually experience things.

Three-Phase Learning Cycle (à la Piaget, 1973)

Phase 1—exploratory hands-on learning, in which children explore freely and ask questions.

Phase 2—idea development, in which students invent concepts and rules that help them answer their questions and sort out their ideas.

Phase 3—concept application, in which children try out their new ideas by applying them to new situations that are relevant and meaningful to them.

In spite of the early efforts of Dewey, Piaget, and others, until the 1950s the science curriculum focused mainly on content. In 1957, Sputnik, the first orbiting space satellite, was launched by the Soviet Union. This alarmed the United States and they immediately called for changes in science education. U.S. science educators began to move beyond content facts to include process. This included having students think, inquire, collaborate, communicate, and work more like actual scientists. The science inquiry (process) approach really got off the ground in the 1960s. In 1996 and 1997, this was confirmed and updated by the National Science Education Standards (NRC, 1995). Though revised and improved, this carefully studied method has proved itself to be more vital today than ever.

INTRODUCING THE NATIONAL SCIENCE EDUCATION STANDARDS

The National Standards for K through 12 science education focuses on four general content categories: science as inquiry, science subject matter, scientific connections, and science and human affairs. The National Academy Press (NRC, 1995) published the National Science Education Standards report at the end of 1995. By the spring of 1996, it had been distributed to school districts across the country. The National Science Teachers' Association (1997) publishes related material. *NSTA Pathways to the Science Standards*, is a set of three practical guidebooks for helping teachers at all grade levels put the standards into practice (both the standards and the guidebooks can be ordered by calling 800–722–NSTA). This book comments on the standards and is designed to support this shared vision of the future. However, we make no effort to replicate the 260-page standards document or the guidebooks.

The following principles guide the National Science Education Standards:

- Learning is an active process.
- Science is for all students.

- School science reflects the cultural and intellectual traditions that characterize the practice of contemporary science.
- Improving science education is part of a systemic education reform effort.

The world is filling up with the technological offspring of scientific inquiry. Dealing with these new realities requires scientific information for public discussions and the choices that we all have to make on an ever more frequent basis. Whether it is democratic decision making, personal fulfillment, new workplace requirements, or the excitement that comes from understanding the natural world, being able to use processes of science is recognized as increasingly important. The standards imply both excellence and equity when they point out that all of our students must develop scientific knowledge and skills.

What is needed for the United States to have a scientifically capable citizenry? The standards emphasize skilled professional teachers, adequate classroom time, a rich array of materials, accommodating work spaces, active inquiry-based methods, and making use of community resources. Like this book, the standards do not suggest a specific curriculum. Rather, they point to ways of bringing out the best in many different curricula. The goal is to move beyond the constraints of the present and bring coherence to the improvement of science education.

Science Content Standards

The content standards highlight what students should know, understand, and be able to do. Paraphrased as follows, they indicate that students should be able to do the following:

- Become aware of physical, life, earth, and space science through activity-based learning.
- Connect the concepts and processes in science.
- Use science as inquiry.
- Understand the relationship of science and technology.
- Use science understandings to design solutions to problems.
- Identify with the history and nature of science through readings, discussions, observations, and written communications.
- View and practice science in personal and social perspectives.

Successful reform of science education requires addressing issues like the instructional support system, teacher education, assessment, and the school culture. The content standards are crucial and comprehensive. They go beyond content and performance areas found in the math standards to address most of the major issues that influence science education. There are professional development standards, assessment standards, system standards, and more. While not wanting to repeat what you can read in the report, we do feel that the focus of the program standards are so important that we should provide a brief overview.

The focus of the Science Education Program Standards is on the following:

- Providing a consistent science program with the other standards and across grade levels.
- Coordinating the science program with mathematics education.
- Availability of appropriate and sufficient resources for all students.
- Providing equitable opportunities for all students to learn about the topics suggested in the standards.
- Developing communities that encourage, support, and sustain teachers.

The program standards call for the inclusion of all content standards across disciplines that are appropriate, interesting, relevant to students' lives, organized around inquiry, and connected to other school subjects. As teachers implement the standards, they will engage students in learning science by actively involving them in inquiry that establishes a knowledge base and connects to the real world in an interesting way. Teachers and students will jointly be part of a learning community that cares about each of its members. The National Science Education Standards make a real contribution to educational reform by providing a map that charts the way to scientific literacy. It is up to us to implement these standards and take science education toward a shared vision of the future.

TODAY'S INTERDISCIPLINARY SCIENCE EDUCATION

When it is taught as an active hands-on subject, science can be an exciting experience for elementary students and teachers By connecting to other disciplines, science can also provide teachers with many opportunities for integration with other subjects. Teachers need subject-matter knowledge that is deep and flexible enough to work with the current influx of second-language learners and students who must face the grinding poverty found in many American cities. This often requires optimizing language and literacy development possibilities to get at content. It also requires reading to gain insight into others' experiences and the way others may be oppressed.

To use and understand science today requires an awareness of how the scientific endeavor makes use of closely related domains (like language and technology) and how it relates to our culture and our lives.

The best science teachers are usually those who have built up their science knowledge base and developed a repertoire of current pedagogical techniques. Many skilled science teachers now begin by making connections between science, language arts, technology, and real-world concerns of the type that might be found in a good newspaper. By stressing real investigations and participatory learning, these teachers move children from the concrete to the abstract as they explore the major conceptual themes that run through science, language arts, and technology.

Teaching strategies in elementary and middle school science now include plenty of participatory experiences and opportunities for students to explore science in their lives. The emphasis on thinking skills, work teams, and inquiry involves posing questions, making observations, reading, planning investigations, experimenting, proposing explanations, and communicating the results. By developing effective interpersonal skills, students can work together to frame questions and to critically examine data. This sometimes means designing and conducting real experiments that carry their thinking beyond the classroom. As language-arts and science instruction becomes more connected to children's lives, enriching questions arise from inquiring about real-world concerns.

ENGAGING STUDENTS WITH THE INQUIRY PROCESS SKILLS

The inquiry skills of language arts, science, and technology (often referred to as inquiry processes) are acquired through a questioning process. As we discussed in Chapter 1, this question about inquiry directs the searcher to knowledge, whether newly discovered by the individual or new ideas not explored before in the field. Inquiry also raises new questions and directions for examination. The findings may kindle ideas, suggest connections, or generate ways of expressing concepts and interrelationships more clearly. The process of inquiry helps the inquirer grow in content knowledge and the processes and skills of the search. It also invites learners to explore anything that interests them. Whatever the problem, subject, or issue, any inquiry that is done with enthusiasm and with care will use some of the same thinking processes that are used by scholars who are searching for new knowledge in their field of study.

As the National Science Education Standards (1995) suggest, the scientific concepts required for a solid foundation in these subjects include observing, comparing, classifying, sequencing, communicating, and solving problems. These concepts develop as children recognize relationships between objects and sets and later as they develop number concepts. These concepts are not exclusive to science, but are necessary in all subject areas. Teaching in an interdisciplinary language-arts and science curriculum recognizes the concept areas mentioned in the science content area but adds and enhances them in what we call the inquiry process skills of language arts and science. We invite readers to take a closer look at the inquiry processes—inquiry skills that enable them to construct their own knowledge. These abilities serve as the tools by which students can acquire additional knowledge in the future.

> *To develop science inquiry skills, K–4 students should be able to plan and conduct a simple investigation, use simple equipment and tools to gather data, form reasonable explanations based on the data they collected, and communicate the results of their findings.*
> —NSTA (1997)

The National Science Education Standards are beginning to have an impact. For example, until recently teaching students the science process skills was considered essential. The standards suggest that we should put less emphasis on the individual process skills and put more emphasis on using multiple process skills (with a purpose in mind). Teachers around the country are now giving students more opportunities to simultaneously explore skills like observing, sorting, describing, and recording.

Poetry can shed light on the power of observation in understanding the natural world.

The Summer Day
from *House of Light*, by Mary Oliver (1992)

> *Who made the world?*
> *Who made the swan, and the black bear?*
> *Who made the grasshopper?*
> *This grasshopper, I mean—*
> *the one who has flung herself out of the grass,*
> *the one who is eating sugar out of my hand,*
> *who is moving her jaws back and forth instead of up and down—*
> *who is gazing around with her enormous and complicated eyes.*
> *Now she lifts her pale forearms and thoroughly washes her face.*
> *Now she snaps her wings open, and floats away.*
> *I don't know exactly what a prayer is.*
> *I do know how to pay attention, how to fall down*
> *into the grass, how to kneel down in the grass,*
> *how to be idle and blessed, how to stroll through the fields,*
> *which is what I have been doing all day.*

Tell me, what else should I have done?
Doesn't everything die at last, and too soon?
Tell me, what is it you plan to do
with your one wild and precious life?

THE INQUIRY PROCESS SKILLS

Observing

The most important tool for young children is observation. Wanting to find out about the natural world makes students eager to explore and ask questions. Observing involves using all of the senses—seeing, hearing, tasting, smelling, and feeling—working together to gather as much information as possible. It is an immediate reaction to one's environment. Students should be directed to describe what they see, hear, smell, touch, and perhaps taste. Encourage students to try to give some specific measurements to their observations. Warn students not to allow their observations to by influenced by preconceptions. Even young children should be able to make good observations.

Sorting and Classifying

Students learn about objects by grouping and ordering them. Classifying relies primarily on observation. As children become more skilled in recognizing characteristics of objects, they learn to recognize likenesses and differences between objects. Classification is an important part of our lives. Shopping at the supermarket, finding a book in the library, or even setting the dinner table would be a tremendous time-consuming chore if things were not classified. At a young age, children are able to classify or sort objects into groups by color, size, or shape, rearrange the set, and put the groups in some kind of order.

Comparing

Once children learn to observe and describe objects, they soon begin to compare two or more objects. Young children may say they want more or fewer, and they can tell you what is the same or different. Being able to compare individual and sets of objects will help children decide whether four is more or less than six. Comparing is not just a skill for students in the early grades. Good science involves comparing other studies, experimenting, and reaching conclusions. Teachers should offer a variety of activities that let children use all of their senses, group objects in many ways, and encourage students to interact with others and communicate their findings.

Sequencing

Children live with sequences and patterns. They may notice patterns in nature (the symmetry of a leaf or the wings of an insect) or patterns in the classroom (the tessellations of the floor or ceiling tiles). Sequencing is finding or bringing order to their observations. These interesting patterns all around are enlivened when teachers direct student observation and pattern finding. Watch children as they put a variety of objects in order: Do their groups have a common attribute? Are the objects arranged in a particular way? Have students explain their groupings and reveal clues about their science skill understanding.

Measuring

At the middle elementary level, students are able to master skills of a good inquirer. Many students make measurements using different tools: rulers, thermometers, scales, clocks, and so on. Active experiences in both science and the language arts provide opportunities to describe and compare in terms of quantity. Young children automatically use descriptive language when comparing quantities (one child is taller than another, one book is heavier, one ball is larger, etc.). In both science and language arts, researchers are constantly measuring. Measuring supplies the hard data necessary to confirm hypotheses and make predictions. It provides first-hand information. Measuring includes gathering data on size, weight, and quantity. Measurement tools and skills have a variety of uses in everyday adult life. Being able to measure connects language and science to the environment. Measurement tools give children opportunities for interdisciplinary learning in subjects such as social studies, technology, art, and music.

Communicating about Science

Communication stresses the importance of being able talk about, write about, describe, and explain science ideas. Symbolism, along with visual aids such as charts and graphs, should become ways of expressing language-arts and scientific ideas to others. This means that students should learn not only to interpret the language of science, but to use that language, both in and beyond the classroom. All students gain by regularly talking, writing, drawing, graphing, and using symbols, numbers, and tables to help them think and communicate their ideas. By making sense to others, they indirectly convey the concept in a meaningful way for themselves.

Scientific language and everyday language are interrelated. We need to get rid of the idea that scientific thinking and language are somehow removed from everyday experiences. They overlap and reinforce one another. If students practice thinking and communicating, they will connect their day-to-day experiences and discover a wealth of new techniques.

Learning to effectively communicate makes the world of science outside of school more accessible. It also promotes interaction and investigation of ideas within the classroom as students learn in an active, verbal environment. Working with language arts and science to communicate benefits all students, whether they are fluid in English or another language or are developing proficiency. Children need to experience the precision of speech and writing that good science demands. Communication is most effective when students express themselves with a purpose and audience.

Using Data

The disciplines of science and language arts identify statistics and probability as important links to many content areas. The skills of data gathering, analyzing, recording, using tables, and reading graphs provide many opportunities for representing, interpreting, and recording that apply to many science and language concepts and skills.

Many decisions are based on market research and sales projections. If these data are to be understood and used, all people should be able to process such information efficiently. For example, consider the science and literacy concepts involved in the following:

- Weather reports (decimals, percents, probability, observing weather patterns, classifying climate zones, identifying weather fronts).
- Public opinion polls (sampling techniques and errors of measurement).
- Advertising claims (hypothesis testing).
- Monthly government reports involving unemployment, inflation, and energy supplies (percentages, prediction, and extrapolation).

All the media depend on techniques for summarizing information. Radio, television, and newspapers bombard us with statistical information. The current demand for information-processsing skills continues to grow.

Graphing

Graphing skills include constructing and reading graphs, as well as interpreting graphical information. They should be introduced in early grades. The data should depend on children's interest and maturity. Here are a few kinds of survey data that could be collected in the classroom:

- Physical characteristics (heights, eye color, shoe sizes).
- Sociological characteristics (birthdays, number in family).
- Personal preferences (favorite television shows, favorite books, favorite sports, favorite food).

Each of these concepts give students the opportunity to collect data themselves.

Graphs are an important form of communication in mathematics and science. Graphic messages can provide large amounts of information at a glance. Often graphs are used to make predictions. There are two rules to remember about graphing.

First, there are two types of variables on a graph. A manipulated variable, or it may also be referred to as an independent variable, is always plotted along the horizontal axis. The variable that is plotted along the vertical axis is described as the responding variable or dependent variable. Normally, when the term manipulated variable is used it describes the data that are controlled by the experimenter. For example, a student group had predicted that a majority of the class would be registered as Democrats in the last election. Their question was, "What political party do you belong to and who did you vote for president in the 1996 election?" After they gathered their data through a survey, they organized the data and graphed the results. The manipulated variables in this case were the political parties, the responding variables were the candidates for president.

Second, in creating a graph it is important to make the graph large and clear enough for the reader to make interpretations, predictions, or analyses (Heddens & Speer, 1994).

> *Competence in using process skills provides children with the ability to apply knowledge, not only to science, but outside the classroom in everyday life.*
> —Mechling & Oliver (1983)

Communications Skills

Whether it is listening, speaking, writing, reading, or using communications technology, language is a window into students' thinking and understanding. Oral language is one means of communication. Let us use it as an example. One of the goals in the classroom is to facilitate the use of oral language and listening as a means of communication and learning. Talking is very important in the language-arts and science classroom. Language is central to the quality of the students' scientific inquiry, and listening to students' language is a valuable way to get feedback from students' efforts. There are many ways to give students opportunities to practice and use language effectively. Effective communication will, of course, depend on topical knowledge, but also on students being aware of how to go about communicating orally. Self-awareness also fits in—how well one is applying their oral communication skills. There are many different ways to involve students actively in this process: storytelling, directed reading activities, art (clay modeling, drawing, sculpture, etc.), music (playing, singing, listening), oral presentations, small-group discussions, and creative dramatics.

Sharing

The process of sharing helps students feel more comfortable and less inhibited in speaking before an audience. It builds self-confidence. By sharing their work and ideas, students develop independence (Manzo & Manzo, 1997). Many times, whole-class discussions are held after the children have had time to explore a particular activity or idea. Often, teachers use these group sharing times to summarize and interpret data from explorations. Group sharing is a time for students to discuss their ideas, focus on science and language relationships, and make connections among activities.

Using Space–Time Relationships

As children compare, classify, and sequence objects, they soon look at relationships among objects. Relationships are rules or agreements used to associate one or more objects or concepts with another. Science is a collection of relationships among objects or concepts. A concept in nature is that animals have certain needs: air, food, water, and space. A variety of factors affect the ability of wildlife (animals) to maintain their survival over time. The most fundamental of life's necessities are the needs mentioned. Everything in natural systems is interrelated. If one of these needs was eliminated, the animal population would dwindle and die.

For teachers of language arts and science, the fact that science concepts are relationships is very important. It is almost impossible to show children an example of a science concept without having them compare it or show a relationship to something else. It is critical that we get students to be active mentally, and to reflect on things presented in class. That is the way that the mind can construct a relationship (Van de Walle, 1997).

Predicting

Will it snow tomorrow? How long will I remain in San Francisco? Is my car going to last through the semester? These questions can be answered, but they can only be answered by a guess, not a prediction. They require speculation. How can we turn speculation (guesses) into predictions? For example, the answer to the question about will it snow tomorrow can be highly speculative. But observers can turn that speculation into a prediction. First, students must know something about current weather conditions: How fast are weather systems moving? From what direction are they moving? What is the temperature, relative humidity, dew point? What kinds of clouds are there? Last year at this time I remember it snowed. Are there any low-pressure systems nearby? What about cold or warm fronts? With these data readily available, students can check past weather conditions. If in the past when conditions were similar it snowed 50 percent of the time, there is a 50/50 chance that it will snow tomorrow too: Predictions tell us that, given these conditions, it has snowed every time, sometimes, or perhaps never.

Children learn that not all predictions are accurate. Often there is a high degree of uncertainty in predicting. The ability to make predictions is based on skillful observation, inference, quantification, and communication. Students who understand predicting are aware that unforeseen events can change the conditions of a prediction, and that 100-percent accuracy is unlikely.

Estimating

The curriculum should include estimation so students can explore estimation strategies, recognize when an estimation is appropriate, determine the reasonableness of results, and apply estimation in working with quantities, measurement, computation, and problem solving.

Not too many years ago, to compute meant almost exclusively one thing: pencil-and-paper computation. Today there are many more options, and a computation by hand is usually a last resort. Let us start at the beginning. First, look at the data. Does the data require an exact answer or will an approximate answer do? If an approximate answer will suffice, an estimate can usually be arrived at mentally.

If an exact answer is required, we frequently try to work the computation out in our minds (mental computation). Mental computation is probably a more realistic option than you may suspect. However, due to the numbers involved, a person may have to use other options: a calculator, computer, or work the problem out by hand. A calculator is almost always the choice, unless there are many similar computations; where computations are related to formulas, then a computer is called upon.

If mental computation was always possible or just as fast as making an estimate, there would be little need for estimation skills. The goal of estimation is to quickly determine a result that is adequately accurate for the situation. In everyday life, estimation skills prove to be tremendously valuable and time saving. Teaching these skills to students has become more important in recent years.

Inferring

The basic process skill of inference involves making conclusions based on reasoning. Children often make guesses or inferences about their observations. Inferences are based on observations and experiences (if the observations are inaccurate or flawed, inferences will reflect this). Students are often very creative in making inferences based on what they have observed. Inferences extend observation by allowing learners to explain their findings and predict what they think will happen. Are these inferences correct? All anyone can say is that inferences either relate to the observations in a logical way and are therefore reasonable, or they do not (Neuman, 1993). The interesting and challenging thing about making inferences is the language that takes place among the students. Language is a powerful tool for gathering and sharing information. Children should therefore be encouraged to talk to each other, the teacher, and other adults while they are engaged in observing and making inferences based on their senses and experiences (Hamm & Adams, 1992).

EXPERIMENTING

The basic inquiry processes just discussed are global in their application, and not limited only to science. For example, students might use the process of inferring to try to understand why their teacher was angry with them in the lab yesterday. A student might sort and classify her supplies for the field trip tomorrow. The experimental inquiry processes that will be looked at in this section are more limited in their scope and application.

Inquiry processes, such as forming an hypothesis, identifying and controlling variables, or analyzing data are skills used in a controlled experiment. Let us look closer at the experimental process skills.

The following are points to remember about forming hypotheses:

Hypotheses are not merely educated guesses.

Hypotheses are testable.

Hypotheses state a relationship between three variables:

A manipulated or independent variable (this is the variable that the experimenter adjusts or alters).

A responding or dependent variable (this is the variable that is influenced or changed by the variable that the experimenter manipulates).

Intervening variables (these are all the other factors that influence the relationship between the independent and dependent variables).

ACTIVITIES USING INQUIRY SKILLS

Activity 1: Buttons and Shells

Inquiry Skills: observing, classifying, comparing, sequencing, solving problems, group work, communicating, recording, gathering data, comparing, measuring

New Vocabulary: sorting, seriation, gradation, Venn diagram

Materials: bags of assorted buttons (one for each group); yarn or string dividers, bags of small seashells (one for each group)

Problem 1: How many ways can you sort your bag of buttons? Try to sort them at least ten different ways.

Problem 2: Make a Venn Diagram using your buttons.

Venn Diagram: a method of illustrating set unions and intersections. For example, a set of blue buttons is one category, a set of round buttons is another category, a set of blue round buttons is an intersecting category.

Problem 3: Classify the shells by seriation: light to dark (color), small to large, number of ridges, patterns, amount of water shell can hold.

Activity 2: Mystery Liquids Experiment

Inquiry Skills: hypothesizing, experimenting, and communicating

Activity: In this exploring activity, students are experimenting with chemicals and doing physical science work. They are learning to use tools found in the lab and becoming familiar with the safety rules of science, mathematics, and technology.

Problem: Try to find out what the four mystery liquids are.

Rules:
1. Each liquid is a household substance that may or may not have been colored with food coloring to hide its identity.
2. Students are limited to using only their sense of sight to do this experiment. Students should experiment by manipulating.
3. For safety reasons, caution students they are not to smell, touch, or taste the chemicals.
4. Each medicine dropper must be used to pick up only one liquid. We do not want contamination!

Problem: Try to find out what the four mystery liquids are.
1. Write a description of the activity.
2. Explain how the group went about solving the problem.
3. Write what the group learned from the activity.
4. Give some follow-up suggestions of how the lesson could be improved or provide suggestions that would follow the activity.

Activity 3: Demonstrating the Behavior of Molecules

Inquiry Skills: observing, comparing, communicating

Background Information: In this demonstration, we will discover more about energy. Molecules and atoms are the building blocks of matter. Heat and cold energy can help change molecular form.

These are questions children naturally ask: "Why does ice cream melt?" "Why does the tea kettle burn my hand?" "Where does steam come from?" "Why is it so difficult to break rocks?" We need to assess students' understanding of the concepts. Students enjoy participating in a hands-on demonstration of how molecules work. Before beginning the demonstration, we need to know that matter and energy exist and can be changed, but not created or destroyed.

Student volunteers are asked to role play the parts of molecules. Direct the volunteers to join hands, showing how molecules are connected to each other, explaining that these represented matter in a *solid* form. Next ask them to show what happens when a solid becomes a *liquid*. Heat causes the molecules to move more rapidly, so they no longer can hold together. Students should drop hands and start to wiggle and move around. The next question will be, "How do you think molecules act when they become a *gas*?" Carefully move students to the generalization that heat transforms solids into liquids and then into gases. The class will enjoy watching the other students wiggle and fly around as they assumed the role of molecules turning into a gas. The last part of the demonstration is the idea that when an object is *frozen*, the molecules have stopped moving altogether. The demonstration and follow-up questions should spark a lot of discussion and more questions.

Activity 4: Popcorn Activity

Inquiry Process Skills: hypothesizing, observing, predicting, estimating, measuring, communicating, collecting data

Materials: butcher paper, air popper, popcorn, rulers

Background information:

What causes popcorn to pop?

Popcorn is a favorite snack of millions, small hard kernels of dried corn that explode into a fluffy tasty treat. What makes that happen? Listen to the noise in the popcorn popper. You will hear the tiny grains of corn popping open. They are turning inside out. The corn looks different now. It is big and fluffy and white.

What puts the "pop" in popcorn?

A little bit of water is the magic ingredient that puts the pop into popcorn. In order to pop properly, a kernel should contain about 13.5 percent water. When the popcorn is heated to 100 degrees C (212 degrees F) the water changes to steam. Trapped inside the kernel of corn, the steam pushes outward, trying to escape as the temperature rises even higher. At about 200 degrees C there is so much pressure that even the tough kernel can no longer hold the steam inside. The corn explodes with the familiar pop. It puffs up thirty to thirty-five times its original size. If you weigh popcorn before it pops, you will find it weighs more than after it explodes, even though popping increases its size about 3,000 percent or more. Can you figure out why?

What makes the popcorn pop?

Heat is what makes popcorn pop. Fire makes the corn hot. The inside swells up. It gets bigger. Soon the inside is too big for the outside kernel cover. So the corn seed pops open.

Historical facts about popcorn

Long ago colonists served popped corn with cream for breakfast.

Columbus saw Native Americans wearing popcorn as jewelry.

Popped corn 5,000 years old was found in a bat cave.

Popcorn was brought to the first Thanksgiving dinner.

Reader's Theatre: Connecting Science to Children's Literature

Next, it is time for students to become actively involved in a group reading of a children's story. For older students, you may direct them to present this poem and reader's theatre to a primary classroom following the directions listed.

Directions for this reader's theatre activity

Kneel down, when your turn to read comes, hop up and read your part. When not reading, kneel back down (or use chairs—stand when reading, sit when listening). Groups can write these stories from anything that they are reading or they can use scripts already prepared. Groups practice in their small group first and do their reading in front of the whole class (if you have five children in a group and eight roles, some students get to read two).

The Song of the Popcorn (first-grade level story)

> Everyone: Pop, pop, pop!
> 1st child: Says the popcorn in the pan!
> Everyone: Pop, pop, pop!
> 2nd child: You may catch me if you can!
> Everyone: Pop, pop, pop!
> 3rd child: Says each kernel hard and yellow!
> Everyone: Pop, pop, pop!
> 4th child: I'm a dancing little fellow!
> Everyone: Pop, pop, pop!
> 5th child: How I scamper through the heat!
> Everyone: Pop, pop, pop!
> 6th child: You will find me good to eat!
> Everyone: Pop, pop, pop!
> 7th child: I can whirl and skip and hop!
> Everyone: Pop, pop, pop, pop!
> Pop, pop, pop!

Organizing an interdisciplinary lesson around a theme can excite and motivate all students to actively carry out projects and tasks in their group. Additional activities that connect to science might include getting an air popcorn popper, estimating how far popping kernel might fly if the top is off, and letting a dozen pop out. Observe the trajectory and chart the arc of the popping corn. Measure and compare with estimation. Discuss the physics behind this phenomena. Students could also use the Internet to get references about the physical processes involved. You might want to put down some butcher paper or bring some butter and make enough to eat.

Activity 5: Experimenting with Paper Airplanes (Blackburn & Lammers, 1996)

Description: Teachers can turn classroom distractions into a project of design and discovery. In this activity, students will discover that there is more than just folding and tossing paper. As they work on perfecting design plans, they will learn to hypothesize, experiment, and draw conclusions (Clemens, 1996). Students will work in small groups to design a paper airplane of any size using any or all of the materials provided. Their challenge is to design a plane that will fly farther and straighter than the planes built by the other groups. The process skills of predicting, estimating, experimenting (forming hypotheses, identifying variables, collecting data, analyzing, and explaining outcomes) are applied.

Materials:
    six or seven grades of paper: typing, onion skin, computer paper, construction paper, paper towels, cardboard, milk cartons

    paper clips in various sizes, staples, tape

    directions for designing paper airplanes

Objectives:
1. Students will design a plan for their airplane.
2. Students will formulate a hypothesis describing their design and their projection of a successful flight pattern.
3. Students will experiment with the materials and modify their design.
4. Students will identify the variables that influenced the outcome of their investigation and record their efforts.
5. Students will carry out the investigation and generate data.
6. Students will communicate their data through written procedures.
7. Students will actively participate in the plane throwing contest.

Procedures:
1. Give each group a copy of the paper-airplane directions and at least three sheets of paper.

2. Introduce the class to some factors that can affect the performance of paper airplanes.

Folding: Symmetry and sharp folds are crucial in designing the plane.

Adding weight: A paper airplane needs weight at the front tip (nose). In many cases, the folded paper provides the weight, but if the nose is not heavy enough, the plane will rise up in front, then fall straight down. Paper clips, staples, tape, or additional folds can add weight. If a plane is too heavy it will dive to the ground. To give it more lift, cut and fold flaps on the backs of the wings. If the flaps are folded at 90-degree angles, the plane will fly differently than if they are only slightly turned up.

3. Encourage students to experiment as they adjust the variables. They will learn a lot about trial and error as well as making and testing hypotheses.
4. Groups test their designs and try out their model experiment. Allow lots of time for practice.
5. Before the contest begins, the class may wish to design posters (stating their purpose and the skills involved) and invite other classes to watch their science and math airplane contest.
6. As a class, conduct the airplane contest. Airplanes will be judged on how far they fly and how long they stay in the air (use a stopwatch). If students create designs that loop in flight, they may also want to judge the number of circles. The best place to hold the contest is in the school auditorium (no wind, plenty of space). Allow each group to fly their model two or three times, then take the best score.
7. Groups should present their model, explaining their hypotheses and how it was assembled.

Evaluation:

Students have the opportunity to ask questions and share designs and launching tips with their classmates. Students write their reflections and feelings about the project (frustration, satisfaction) in their notebook or portfolio.

Activity 6: Bubble Making (Zubrowski, 1979)

Description: Everywhere people of all ages and backgrounds create soap bubbles. A soap bubble is formed when a thin film forms a sphere or hemisphere around air. The most exciting bubbles are the ones we blow for enjoyment outdoors on beautiful days through bubble wands or pipes. The most common bubbles are those we see floating in the sink or bathtub. This activity gets students actively engrossed in the process skills of exploring shapes and symmetry in nature.

Inquiry Skills: recording data, sharing techniques, drawing diagrams, discovering and recording patterns

Materials:
- plastic coffee stirrers or milk straw (half the length and diameter of a standard plastic soda straw)
- paper towels (one per group)
- one cup or container per group, large enough to hold an ounce
- bubble solution (mix one cup of water with four tablespoons of liquid dishwashing soap)
- masking tape and pen

Objectives:
1. Students will explore shapes and patterns of bubbles.
2. Students will share the excitement of bubble blowing with others.
3. Students will draw diagrams and take notes of all the bubble arrangements they find.
4. Working in pairs, students will compare and investigate the symmetry of the bubbles they blow.

Procedure:
1. Pour bubble solution into cups.
2. Wrap a length of tape around one end of the straw. It should look like a flag.
3. Each person writes their name on the tape (the tag is a reminder of which end of the straw to put in your mouth and identifies each person's straw).
4. Wet the paper towels.
5. Working in pairs, students blow hemisphere-shaped bubbles and record the patterns the bubbles make on their hand.
6. Wet one palm holding the fingers together, so the hand can be a table for their bubble to rest on.
7. Dip the end of the straw into the cup of bubble solution.
8. Hold the straw into the cup of bubble solution.
9. A bubble hemisphere should begin to form on the student's palm.
10. Students should be able to feel the air that they blow inside the bubble on the palm of their hand. The size of the bubble increases when air is added.
11. Have students practice a few times.
12. Working only in a single layer, have students draw diagrams of all different arrangements of bubbles they find. Encourage them to take notes on how to duplicate the arrangements they discover.

Evaluation:
Review the shapes and arrangements each group made to see the natural patterns made by the class. The class may wish to design a graph or make a poster of their hemispheric investigation to share with other classes. A written account is also helpful.

REVIEWING THE PROCESS SKILLS

Science and mathematics inquiry processes are tools that enable students to gather and discover data for themselves. To summarize, some of the major skills are observing, classifying, inferring, measuring, comparing, sequencing, communicating, predicting, recording, investigating, and experimenting. To teach the inquiry processes, students investigate subject matter with concrete materials. Their thinking can be guided with questions.

In observing, students learn to use all of their senses, note similarities and differences in objects, and be aware of change. In classifying, students group things by properties or functions, and they may also arrange them in some sense of order. Sequencing is part of this ordering system. Measuring teaches students to find or estimate quantity. Measurement is often applied in combination with skills in an integrated mathematics program. Communicating involves students in organizing information in some clear form that other people can understand. Recording, graphing, and using maps, tables, and charts contributes to the communication process. The skill of inferring requires students to interpret or explain their observations. When students infer from data what they think will happen, the term predicting is often used. The most challenging process, one that usually takes place with students in fourth grade and up, is experimentation. This process is divided into the following subskills: forming hypotheses, identifying variables, collecting data, analyzing data, and explaining outcomes.

Science education now goes beyond life, earth, and physical science. Teaching a concept in depth is more important than "coverage." Less, it turns out, really is more. The research and literature suggest focusing on the ideas and skills that have the greatest scientific and educational significance. It may take more time, but children learn by doing, reflecting, communicating, and working in a caring community of learners.

THE PROCESS OF SCIENTIFIC INQUIRY

Most science educators today agree that science can best be viewed as a continuous process of trying to discover order in nature and looking for consistent patterns of the universe through systematic study. The question is the cornerstone of investigation. It guides the inquirer to a variety of sources, revealing previously undetected patterns. These undiscovered openings can become sources of new questions that can deepen and enhance inquiry. Science is a way of thinking and asking questions about the workings of the universe, and the scientific method (process) is an intellectual tool that is just as important as the tools of analog for history or metaphor for English. Though the belief that the process of scientific change is strictly rational was punctured years ago, scientists do share basic beliefs and reasoning processes as they go about their work. It may not be true of Nobel Prize winners, but the typical scientist tends to accept what he or she was taught and applies his or her knowledge to solving the problems that come before him or her.

As Thomas Kuhn has pointed out, conventional scientists often accept a paradigm (like Ptolemy's theory that the sun revolves around the Earth) and then go about solving problems in ways that do not disturb the basic paradigm. It may take childlike openness—and sometimes true genius—to do research that contradicts the theories already in place. In fact, it often takes a young scientist who is not deeply indoctrinated in the accepted theory (like a Lavoisier or an Einstein) to sweep the old beliefs away.

Children are more likely to be open to new approaches and have a natural curiosity when it comes to inquiring about the world. Even very young children learn by experiencing things for themselves, thinking uniquely, and communicating what they have done to others. Scientific inquiry can help children's curiosity to blossom. Questions have been asked by children throughout history. Clearly, some of their answers were wrong, but the important thing is that children never stopped asking: They saw and wondered, and they sought an answer.

THOUGHTFUL QUESTIONING

The right question at the right time can move children to peaks in their thinking that result in significant steps forward and real intellectual excitement.
—Eleanor Duckworth (1987)

Have you ever wondered about nature and about how things work? What questions do you want to know the answer to? An example of ways of thinking and asking questions is an activity like the following:

- Think of a science question you are curious about.
- Get together with other classmates and compare questions. As a group, come up with one good question. Form a hypothesis.
- Write the group question in your portfolio.
- Write your group's hypothesis.
- Do some research (consult books, perform an experiment that proves or disproves your hypothesis).
- Present your findings to the class.

Some commonly asked questions are as follows:

- Why does the moon change shape?
- What causes my can of soda to have water droplets on the outside?
- How does an airplane stay up in the air?

One question leads to many more. Helping students develop the ability to inquire is central to science learning. Like scientists who search for knowledge, children are intrigued by the unknown and unexplained. Questioning is basic to societal and scientific progress (Trowbridge & Bybee, 1996).

Science activities often have surprising results. A simple activity uses a ping-pong ball and a funnel to demonstrate pressure. Show the class a ping-pong ball inside a funnel. Set up the problem: Ask how far students think the ping-pong ball will go when they blow into the funnel. What happens? Students will soon discover that as they blow into the funnel the ball does not move. They are proving Bernoulli's principle that when air moves faster across the top surface of a material the pressure of the air pushing down on the top surface is smaller than the pressure of the air pushing up on the bottom surface.

Having a knowledge of science is valuable for everyone because it makes the world more meaningful and more exciting. All students should have an awareness of what science is and how it relates to their culture and their lives. This means understanding how language, science, and technology connect in our daily routines. It also means recognizing the historical base and human contributions. In the final analysis, it is the human dimension that determines the limitations and miracles of science.

SCIENTIFIC MISCONCEPTIONS

Concern has been voiced in recent years about what students are actually learning in science and the influence of scientific falsehoods on the American population. There is general agreement that American children are awash in media-induced misconceptions. It is little wonder that fewer than half of adult Americans have even a rough idea of how long it takes the earth to go around the sun. Worse yet, newspaper surveys report that one in five of us is sure that the sun rotates around the earth. In a 1990 public survey, two in five Americans believed alien creatures have recently visited the earth (Miller, 1990). And students who watch TV shows like "The X–Files" might find it more difficult to look critically at visitations by intelligent extraterrestrial life.

With a steady diet of movies and TV programs like "The X–Files," the unreal becomes believable. The skepticism about abnormal claims that is part of scientific thinking is often viewed as a handicap in learning "what is real." It is better to worry about those strange creatures from outer space. In the last twenty years, science has been portrayed as useless in identifying and solving such problems (Evens, 1996). Pseudoscientific nonsense seems to speak to powerful emotional needs that real science may leave unfulfilled. Wishes and personal fantasies about supernatural powers are often the cornerstones of false science. When we self-indulgently confuse hopes with facts, we slide back into pseudoscience (Sagan, 1995).

Science is more than a body of knowledge, it is a way of thinking. Every time we exercise self-criticism and test our ideas against the outside world, we are doing science. There is nothing wrong with a good science fiction story as long as we do not confuse it with natural reality. Science transports us toward an understanding of how the world is, rather than how we wish it to be. When we pass beyond this barrier, when we understand and put this knowledge to use, when we finally "get it," deep wonders are revealed.

OVERCOMING THE MISCONCEPTIONS BROUGHT INTO THE CLASSROOM

Studies show how children try to reconcile their own experience with what they have been taught, what they have heard, and what they have viewed on television or in the movies. Often teachers deceive themselves into thinking that effective instruction has occurred when students are able to successfully solve items on tests. We know that developing real understanding of science is much more than successfully completing test items. Misconconceptions are rooted in commonsense explanations and trying to make sense of poorly understood scientific descriptions. From horoscopes to lotteries, explanations are invented to fit private understandings. For example, when we ask second graders about the shape of the Earth, most will say that it is round like a ball. When questioned further, the majority usually admit that they believe the Earth is flat. They had been taught in school that the Earth is round like a ball, but their personal experiences tell them that the Earth is flat. A child's writings reveal her thoughts: "The earth has a flat plane through the center, with the sky above and rocks, water and dirt underneath" (Maili & Howe, 1979). One important finding from the research on children's misconceptions is that they may comprehend very little of the science being taught if they have deeply ingrained personal ideas about the subject (Perkins & Simmons, 1988).

Recent studies conducted at prominent universities found that even college graduates had difficulty in defining simple concepts about the phases of the moon, the reason for the seasons, and the causes of the apparent motion of the planets and stars (Schneps, 1988; Gardner, 1991). This material is supposedly covered in elementary grades. These results revealed deep disparities between teaching, testing, and the reality of student understandings, from the elementary school to the college campus (Tinker, 1991). It is little wonder that scientific literacy and teaching for understanding have emerged as important goals for science education. The recognized importance of a scientifically literate citizenry is one of the main reasons for national efforts to reform science education.

Part of the answer to improved scientific literacy lies in teaching for deeper understanding, helping students relate school concepts to everyday life, using concrete examples, and connecting ideas across disciplines. How is it possible to teach for understanding? How can teachers avoid teaching superficially? Piaget (1973) suggested that learners construct their own knowledge by assimilating new experiences in ways that make sense to them. These ideas, applied to individual and group construction of knowledge, are just as powerful today.

TODAY'S LITERACY REQUIRES A QUALITY SCIENCE PROGRAM

Boundaries between traditional subject-matter categories are being softened and connections enhanced. The amount of detail that students are expected to retain today is considerably less than in traditional science courses. Ideas, thinking skills, and collaboration are emphasized rather than specialized vocabulary and memorized procedures. Science topics and concepts are structured in a manner that makes sense to all students, while providing a solid foundation for learning more. Details are treated as enhancing understanding. For example, instead of lecturing about the properties of water, it would be much more engaging for students to solve a problem on water properties. A teacher could ask the class, "How many drops of water do you think would fit on a penny?" As students estimate and experiment, the teacher guides, focuses, challenges, and provides encouragement.

In the new science curriculum, time is devoted to understanding what science means and how science, the language arts, and technology relate to each other and the social system. This includes some important historical data about science and technology. Both have roots going far back into history in every part of the world. Though modern science is only a few centuries old, many scientific tools and concepts can be traced to early Egyptian, Greek, Chinese, and Arabic cultures. Major conceptual themes that span all scientific thinking form the foundation of today's science curricula.

An Interdisciplinary Activity

The following interdisciplinary activity is one we invite all students to try. It gives them a chance to get in touch with their misconceptions, stereotypes, or ideas. A small group works best. You need a large sheet of paper and crayons or markers. The group needs to answer the questions about their scientist and present their scientist to the class.

Draw a Scientist

1. Problem: Present students with a problem called "Draw a scientist."

    Students will work in groups to discuss and show what they think a scientist looks like. They will be required to name their scientist, describe the job of the scientist they drew, tell about their scientist's hobbies, and explain the job functions or responsibilities of their scientist.

    Groups will present their scientist to the class and be willing to answer any questions the class may have concerning their scientist.

2. Evaluation: Encourage students to reflect on the activity.

    Students will write a description of the activity for future students.

    Students will describe what they learned from the activity and offer their ideas and suggestions for follow-up.

Assessment Activity for the Teacher

As you read this book, we suggest that you keep a growth portfolio to record your reactions to the readings, the group activities you tried, and your own explorations. Put in some sketches. In your portfolio, tell about the situations and the ways you have used these processes in your own inquiries or in your teaching.

MAKING SURE SCIENCE HAPPENS

The National Science Education Standards describe a vision and provide a map of what students should understand and be able to do. Many good small-scale federally funded elementary science programs were initiated in the past three decades, but until the 1990s the impact was greater in science education textbooks than it was in actual classrooms. Many teachers were ill prepared to change science instruction. Some tried to avoid the topic altogether. Beginning with the first edition of the math standards in 1989, important new ideas about science instruction started to percolate down and into many American elementary schools. Reforms related to the education of teachers, technologies for teaching, performance assessment, the organization of schooling, and educational research are now having an even greater effect. Providing standards that spell out what is best for students and key areas for intellectual engagement helps. The openness to new approaches inherent in teacher-initiated movements, like whole language reading instruction, also seems to have had a liberating effect on teachers. Fundamental change suddenly becomes a possibility.

Science teaching is changing and progressing as agreement is reached on what children need to know (standards) and as we find out more about how children learn (pedagogy). As elementary school teachers strive to work out the best approach to content standards, it is important to be aware of the information that is coming out about the best way to go about teaching that content. How to teach science well is just as important as the what and why of teaching.

Every question, including, "Why are bubbles kind of round?" has an element of deep mystery. And no matter how hard science tries to represent the world, things often remain just a little out of reach. It is the job of the teacher to help children come to terms with an uncertain universe. Whether it is learning about sound, magnets, plants, popcorn, or bubble blowing, good instructional techniques can make science come alive and get students closer to natural realities. Exploring mysteries through hands-on scientific inquiry is exciting for everybody. Teachers work magic every day, but they cannot move mountains by themselves. All parts of the education system must work together and support teachers. This includes other teachers, administrators, parents, teaching peers, university researchers, the community, business interests, and government agencies.

It is difficult to suggest what children will need to know in the 21st century. Visions of the future can only be speculations and extrapolations from today's rapidly shifting environment. However, some general principles are predictable. For example, it seems a safe bet to say that students will need to be literate, critical thinkers who can use technology and work with others. They will certainly have to be lifelong learners who know how to go about upgrading their skills once their formal schooling is over. And knowing how what they are learning can be applied in the real world is bound to be important for the individual and for society.

> *The scientifically literate person is one who is aware that science, technology, and communication are interdependent human enterprises with strengths and limitations; understands the key concepts and principles of science; is familiar with the natural world and recognizes both its diversity and unity; and uses scientific knowledge and ways of thinking for individual and social purposes.*
> —Rutherford & Ahlgren, 1990

Chapter 6

The Language Arts and Literacy

Children learn language and its uses simultaneously.
—Frank Smith, quoted in NCTE (1996, p. 17)

Language arts is more than reading and writing. It includes the capacity to accomplish a whole range of communication and language tasks. In contemporary society this means being active, critical, and creative users, not only of print and written language but also of spoken language, visual language, and more. Here the emphasis is on an integrated language-arts approach to language learning that connects students to literature and to real communication situations. The visual literacy and information technology sides of the equation are covered more fully in other chapters.

Language is learned holistically and in context, rather than in isolated bits and pieces. Learning from practice exercises is less effective than reading real text or writing for a real audience. From a constructivist perspective, this means building on a child's prior experience, teaching the use of context clues, and helping students make inferences about meaning. Whatever one's philosophical base, there is general agreement that students should frequently read, write, speak, and use communications technology.

In contrast to many basal reading programs, which provide step-by-step guidelines, an integrated language-arts approach calls on teachers to make decisions about what to teach and when to teach it. Informed beginners are welcome. This chapter is designed to assist teachers who are looking for new ways to make the language arts more active, dynamic, purposeful, connected, and fun. The goal is to help new and experienced teachers share in the richness and excitement of helping their students comprehend and apply the language arts.

Language-rich home and school environments are key factors in the development of language and literacy skills. When it comes to technology, tools like the Internet and e-mail make people read and write (or at least type) more, but not necessarily better. Children learn some of their most important language and literacy lessons when they have interesting conversations with adults. This might, for example, take place at the dinner table, when children learn new vocabulary and describe events beyond the here and now. Teachers can help a little with this aspect of literacy by making themselves available for interesting conversations with their students, but they must go on to make sure that children frequently engage in reading, writing, speaking, listening, critical reflection, and collaboration.

INTERNATIONAL READING ASSOCIATION (IRA) AND NCTE STANDARDS FOR THE ENGLISH LANGUAGE ARTS

The following are the IRA/NCTE Standards (NCTE, 1996):

1. Students read a wide range of print and nonprint texts to acquire new information, meet the demands of society, and for personal fulfillment.
2. Students read a wide range of literature from many periods, genres, and cultures to build an understanding of human experience.
3. Students apply a wide range of strategies to comprehend, interpret, and appreciate texts. They draw on prior experience and interaction with other readers and writers.
4. Students adjust their use of spoken, written, and visual language to communicate effectively with a variety of audiences for a variety of purposes.
5. Students employ a wide range of strategies as they write and use different writing process elements.
6. Students apply knowledge of language structure, language conventions, media techniques, and figurative language to create, critique, and discuss print and nonprint texts.
7. Students conduct research on issues and interests by generating ideas and questions, and by posing problems.
8. Students use a variety of technological and informational resources to gather and synthesize information and to create and communicate knowledge.
9. Students participate as knowledgeable, reflective, creative, and critical members of a variety of literacy communities.

10. Students use spoken, written, and visual language to accomplish their own purposes (e.g., for learning, enjoyment, persuasion, and the exchange of information).

This is a selective outline. For a more complete explanation, see the Standards document.

A COLLABORATIVE APPROACH TO LANGUAGE ARTS AND LITERACY

In many language-arts classrooms today, peer collaboration is seen as critical for helping students develop as literate people. For example, reading and writing are meaningful, constructive, human, and language-based processes which can be enhanced by social activity among peers. In reading, students bring their worldviews and experiences to texts to evoke new meanings (Rosenblatt, 1993). Constructing meaning often involves sharing interpretations of texts with peers within literature discussion groups or literature circles. Writing is seen as a recursive process in which learners make successive approximations toward coherent, vivid writing through sharing their work with others in authors' circles and conferences (Graves, 1994). Students can share their reading and writing experiences to construct meanings, refine ideas, set goals, and assess their growth. All these skills are necessary for literate growth.

The integrated language-arts classroom is grounded by a set of beliefs about teaching and learning. One central idea is that children grow as learners when treated like members of the same literacy club (Smith, 1986) rather than as text-bound pupils searching for "correct answers." While part of a community, students are also seen as individuals, developing in unique ways as they transact with the world. This holistic view of learning implies that there are no rigid, linear stages of literacy development. Students experience growth recursively, based on their developmental capacities as learners within a particular environment.

Errors in reading and writing are not viewed as pathologies; rather, they are seen as windows into how children think (Botel & Lytle, 1990). When a student has difficulty learning how to read or write, it is often a sign that the environment needs to be changed to allow that student to take the risks necessary for development. One way of reducing risks is to encourage students to work with each other. Since peers share a similar age and ability status relative to the teacher, they may be inclined to take risks when interpreting a story or sharing advice about writing (Lazar, 1993).

New ideas about the significance of talk and the place of student authority also underlie the emergence of collaborative learning in the language-arts classroom. Democracy in the language-arts classroom means that students share the power and responsibility of teaching others how to read and write. In more traditional classrooms, peer talk is viewed as "insignificant noise," relegated to few and infrequent periods of the school day (Goodlad, 1983). Increasingly, many educators recognize the positive relationship between talk and intellectual growth (Vygotsky, 1962; Wertsch, 1991). Pairing students with each other allows them to engage in exploratory talk, which is a critical component to learning. Structuring these opportunities also means that teachers come to see students as legitimate "teachers." When children work with peers to talk about a book, revise a piece of writing, or assess artifacts in a portfolio, they become accustomed to social interaction and reaching for learning goals together.

MIXED-ABILITY GROUPS

Increasingly, teachers and schools are setting up flexible or heterogeneous groupings in reading and language-arts classrooms. Traditionally, low–middle–high ability groups have been used to help teachers serve the needs of diverse learners in a classroom. Unfortunately, such structuring has often resulted in situations of permanent tracking; students who do not read and write on schedule with their peers were often placed in the bottom reading group and tended to end up in the lowest-ability groups or classes throughout their school years. Students in lower-ability reading groups usually received instruction that emphasized decoding over meaning-making. Undoubtedly, issues of self-esteem, equity, and opportunity surface when children are seen by themselves and others as less able than their peers.

Heterogeneous grouping allows groups of children of diverse abilities to read and write together, based on criteria like as group's mutual interest in a book or the desire of some students to learn a specific writing convention. Teachers who help children structure these groups must be mindful of the status differences between students and the ways in which students communicate in groups. Experiences for helping to make these groups successful can include teaching learners to be responsive to group members or teaching students how to listen to and give specific feedback to group members. Helping students to monitor their own collaborative activity is an essential part of mixed-ability grouping in the reading and language-arts classroom.

Recent innovations in classroom texts and materials have also influenced peer collaboration in the classroom. Educators are currently struggling to bring practice in literacy instruction into harmony with current knowledge by shifting from basal reading systems to literature-based approaches. Research and theory favor the use of whole, meaningful works rather than works excerpted from their original source. Nevertheless, many reading programs in elementary schools still depend on older basal reading systems that focus on discrete skills and the use of contrived stories that have a controlled vocabulary.

As a response to the literature-based trend, some textbook publishers have even developed literature anthologies. Many of these texts include original stories written by famous authors of children's books. Some of the newer anthologies do not include comprehension questions or vocabulary exercises. Instead, many emphasize meaning-making through collaborative learning. The Houghton Mifflin series, *The Literature Experience Series* (Houghton Mifflin, 1994), offers specific suggestions for having students work in collaborative groups to respond to stories through talking or writing.

New assessment practices allow teachers to gather information about the ways peers transact with one another in the classroom. The current emphasis is on assessing reading and writing processes (as opposed to products), self-assessment, and observation (Rhodes & Shanklin, 1993). Many teachers are assessing how students work with others, including peers, as they move through various reading, writing, talking, and listening experiences. Recognizing that literacy development happens between people (Taylor, 1993) and in many instances between peers, teachers are writing descriptive anecdotal notes about the ways students help each other within collaborative groups. Students are also beginning to assess themselves as participants within reading and writing groups. By integrating all language experiences, teachers develop new ways of seeing students and documenting their growth. This can have a profound influence on the teachers' instructional decision making.

Readers make connections with authors as they read and connect with audiences as they compose. Readers can share what they read with others in organized book clubs or through casual conversations. Writers in the world outside of school often collaborate to help the act of writing, insight, and feeling come together more powerfully. Using all the appropriate technological tools when communicating with others is a natural part of reading and writing. To help students recognize the social nature of writing, teachers can encourage them to randomly examine a library shelf, magazine, or scholarly journal to spot how many works are jointly authored. Published authors can also be invited to the classroom to discuss their collaborations with many people who have helped them along composing routes—wives, husbands, children, friends, co-workers, associates, editors.

Many changes in language-arts teaching have evolved over the last few decades and these innovations have shaped reading instruction in many classrooms. Instead of seeing round-robin reading and silent desk work, visitors to many of today's classrooms are likely to hear lively classrooms, filled with the voices of students talking about books and reading to each other. Emerging portraits from these classrooms continue to reflect new successes, spawning renewed enthusiasm for collaborative learning in the language-arts classroom.

> *Reading is like thinking with the mind of a stranger.*
> —Jorge Luis Borges (1996)

NATIONAL STANDARDS AS A CATALYST FOR CHANGE

Competence in all subject areas is dependent on competency in the language arts. As in other core disciplines, standards for classroom instruction and student learning should be based on the best research and most current knowledge about subject matter, learning, and human development (Ravitch, 1995). A major goal is to help teachers establish English language-arts curricula that value language development, literature, composition, reading, and visual communication. How do subject-matter standards fit in? The language-arts standards offer a set of guidelines, defining what students should know and be able to do in the language arts. None of the standards were carved in stone, but they do represent a comprehensible standards document that can be used to support instruction, insure quality assessment, and provide guidance for the future.

There is general support for a holistic, collaborative, integrated approach to teaching the language arts and helping students learn the language skills they need to deal with today's literacy-intensive environment. This includes an understanding of the features of electronic media and being able to comprehend the structural and symbolic conventions across the whole range of communication possibilities. T. S. Eliot used the term "objective correlative" to explain how clusters of events found in media (particularly in literature) can capture emotional experience. He was referring to situations, objects, or a chain of events that serve as a kind of formula for evoking a particular emotion. Today's technology allows us to create objective correlatives for powerfully thinking about a wide range of stories and subjects.

The language-arts standards that relate specifically to reading include suggestions that teachers help students learn to read strategically This means constructing meaning by self-monitoring comprehension: questioning, reviewing, revising, and rereading. Since students must make sense of literature, film, television, and content-area exposition, it is important that they be able to critically analyze and integrate concepts across information sources. When using the Internet, for example, it is important to understand the difference between something found on the *New York Times* website and the website of the *National Enquirer*. Teachers must help students become part of an active community of readers who understand, appreciate, and discuss whatever they read—wherever they read it.

As far as writing is concerned, there are common themes in the literature and research that cannot be avoided. Writing is viewed as a tool for learning across the curriculum and for fictional stories. Students need to understand the conventions of writing and be able to search out a variety of writing sources, composing with words, illustrations, video, and computer-generated graphics. Attention needs to be given to small-group or pair collaboration using the writing process approach: pre-writing, drafting, revising, and editing. Students apply knowledge of good writing practices (accurate spelling, punctuation, and grammar) while creating, critiquing, and discussing printed commmunications. As students become constructive and critical members of a community of writers, they can help each other and add to the collective spirit.

> *Writing fiction is about allowing the reader to enter another dimension. As a species, we need stories to deal with the collective spirit.*
> —Isabel Allende (1995)

LITERACY CIRCLES AND MEANING-MAKING

An integrated language-arts approach that connects with collaborative groups can have a profound effect on children's work in other content areas. Those most competent at reading also tend to be competent when it comes to speaking, writing, and scientific inquiry (Barr, Kamil, Rosenthal, & Pearson, 1993). If children read good writing about content it has a positive effect on their own writing across the curriculum (Burns, Roe, & Ross, 1988). Putting a strong emphasis on reading real literature is the best way to help children develop a love of books. Students need good books that challenge personal perspectives, stimulate thinking, and clarify intellectual relationships. Books let us enter into unexplored territory and explore the range of human possibility. By taking part in literacy circles, students can discuss what they are reading, investigating, and composing. The whole range of language skills can be helped by communicating with peers who provide immediate feedback during reading, analysis, writing, and revising.

Reading is much more than extracting facts; it involves the making of meaning (from the printed word) and connecting with a rich literacy tradition. Helping readers develop means teaching them to bring prior knowledge to stories, anticipate story outcomes, make connections between stories and their own lives, and read critically.

EXPERIENCES BEFORE READING

For students to make meaningful connections with print, they must be given opportunities to bring their own worldviews and experiences to texts. Structuring before-reading experiences is a way teachers activate students' prior knowledge. The purpose of before-reading experiences is to get students to talk or write about what they already know about a particular subject or theme contained in the text. By reflecting on what they already know, students do a better job relating to text and they make better predictions about story events.

Most before-reading experiences lend themselves to collaborative learning. As a practical example, let us use the novel *Julie of the Wolves* by Gean Craighead George (1972) to show the variety of pre-reading activities that can be structured collaboratively. This is a novel about a thirteen-year-old Eskimo girl who communicates with wolves to survive a long journey across the tundra of Alaska. The main character, Miyax, shows great courage during the journey, and she also learns to identify with the Eskimo way of life. A novel for mature readers (fifth grade and older), this story is packed with rich themes: old versus new, nature versus technology, white versus Native American communities, man versus nature.

Prior to reading *Julie of the Wolves*, students need to talk about what they know about the setting (Alaska), the culture (Eskimo), and the main science topic (wolves) if they are to make important connections with the book. They can also talk about important themes in their own lives which relate to those presented in the book. Working together, students can make predictions about the characters or events in the book after reading the story title. The following is a brief list of specific collaborative activities which students can engage in to activate prior knowledge.

1. Partners can tell each other stories about surviving a difficult journey or accomplishing a difficult job.
2. Partners can generate vivid descriptions of life in the Arctic.
3. Partners can brainstorm the meanings of key words like "survival."
4. Partners can tell each other about their own scary experiences being alone.
5. Small groups can discuss how they would survive alone in the Arctic.
6. Small groups can illustrate a giant map of the Arctic.
7. Small groups can generate questions about the book.

8. Small groups can dramatize an experience about the Arctic.
9. Small groups can preview the text illustrations and make predictions.
10. Small groups can discuss what they would bring on a trip to the Arctic.
11. Students can draw on their science knowledge of wolves.

Another literature and science activity is based on *Night of the Twisters* by Ivy Ruckman (1986), a novel for upper elementary and middle school students. This narrative novel tells the story of the tornadoes that devastated the city of Grand Island, Nebraska. Danny, the main character, is entrusted to take care of his baby brother as his parents seek help. He is able to find safety after the storm with the help of his best friend and eventually they find their families. The following are possible activities:

1. After introducing the novel *Night of the Twisters*, encourage the class to form small groups to do some research on tornadoes.
2. Tell students when they have completed reading and discussing *Night of the Twisters* they will role play the parts of Danny and his friend as they search for their parents. Have them show how they would act if approached by a reporter.
3. Inform groups that they will use the information from the novel and their own research knowledge to create an emergency evacuation plan.

These experiences can help students make important science and language connections with books, prior, during, and after reading. Careful monitoring and observation of groups will help teachers determine how particular students participate in reading experiences.

DURING-READING EXPERIENCES

The primary purpose of during-reading experiences is to help readers make meaning as they progress through the text. Readers can read *aesthetically*, by directing their attention to the lived-through images and feelings that they evoke while reading, or they can read primarily to take away information, as in *efferent* reading (Rosenblatt, 1993). Often, particular types of texts will influence a reader's purpose for reading. Literary texts such as poetry, short stories, or mysteries are more likely to inspire aesthetic reading. Conversely, students may read informational texts such as newspapers, recipes, or prescriptions efferently. Other factors though, like teacher expectations, classroom environment, or student intentions, will shape how readers direct their attention during reading, irrespective of the text genre. Any text can be read aesthetically or efferently, and students can read with their attention directed with some combination of ways.

Teachers need to keep aesthetic and efferent reading in mind when they structure collaborative experiences, as students read together, or between moments of independent reading. Traditionally, teachers have emphasized efferent reading by having students read to answer comprehension questions. More recently, teachers have been interested in structuring aesthetic reading experiences by having students share their lived-through thoughts, images, and feelings about a text. Establishing reader response groups or group literature discussions is an ideal way to get students to simply enjoy good literature (Short & Pierce, 1990). Students make personal connections with texts and share these with peers. They can compare different interpretations of texts and justify their views by finding supporting evidence in the text. To get reader response groups running smoothly, teachers must spend time modeling the process.

Generally, reader response conversations begin with personal responses and move toward higher levels of analysis. They often begin by having students write a personal response after they finish reading a passage or chapter. The written response can be informal words, phrases, sentence fragments, or complete sentences which capture what students are thinking and feeling at the moment. Writing allows time for personal reflection and it also prepares all of the students for a speaking turn. Students are then invited to share what they have written. Inevitably, students will have diverse responses to texts and this inspires another level of conversation around comparing and contrasting viewpoints. This often leads to student's defending their interpretations by looking back at the text to find specific descriptions or actions which "prove" their point. Students can also analyze how they arrived at certain conclusions based on their own prior experiences. They may even critique the writing and think of new story variations.

One possible format for helping students write and talk about their interpretations is the inductive reasoning grid, which is shown on page 105. Using this method, students can write about specific evidence from the text so that they can make more informed interpretations.

Collaborative groups are often used to help students refine their interpretations by gathering and considering evidence as they read. In a directed reading–thinking activity (DRTA), students try to predict future story events on the basis of a few clues that they have been given. Students can read either independently or aloud, stopping to predict what might happen in the story and refining their predictions as they gather new evidence through continued reading. Divergent responses are encouraged, though opinions must be justified by finding supporting evidence from the text. After the story, students can demonstrate their understanding by talking to each other about how their predictions changed as they progressed through the text. The goal of DRTA

Inductive Reasoning Grid

Name of character _____

Reading Clues: *Interpretation:*

1. Statements by the character:

_____   _____

_____   _____

2. Character's actions:

_____   _____

_____   _____

3. Character's thinking patterns:

_____   _____

_____   _____

4. What others say about this character:

_____   _____

_____   _____

5. Situations the character becomes involved in:

_____   _____

_____   _____

6. Other important information about the character:

_____   _____

_____   _____

Summary of the character: My generalizations about the character:

_____   _____

_____   _____

is to get students actively searching for evidence to support their predictions so that they can eventually internalize this way of processing texts while reading independently.

Literature circles (Daniels, 1994) are another way of structuring the classroom for independent reading and collaborative work. Students assume particular roles, such as discussion director, literary luminary, connector, and illustrator, while engaging in literature discussions. Prior to these conversations, all group members choose a text and read it (or a portion of it) silently. They then fill out sheets which outline specific duties of each role:

The discussion director develops discussion questions for the group.
The literary illuminator guides the group to revisit some sections of a text.
The connector relates a text to other readings and experiences.
The illustrator creates and shares a visual display of a text.

These are some of the suggested roles for literature circles. There are others which can be added, such as summarizer, vocabulary enricher, or investigator. Note how each role describes how each group member can transact with story content. They do not suggest a particular status as would be the case for roles such as group checker or monitor; these roles tend to pit students against each other because one student assumes the role of evaluator of other group members. Roles like these should be avoided.

Teachers should model all roles and allow initial practice sessions so that children can experiment with roles and ask clarifying questions about them. Ultimately, the aim of literature circles is to give students a foundation for conducting conversations about books with the goal of withdrawing the role sheets when students have internalized a variety of ways to talk about literature with each other.

Some teachers even like to assign different group roles so that students can experience and develop the strategies for dealing with isolated, overly talkative, or nontask-oriented group members. This allows them to practice strategies of good listening, clarifying, or disagreeing in a nonthreatening manner. Smooth and successful peer interactions depend on clearly defined roles for individual group members.

During-reading experiences are also aimed at helping students become more fluent readers. Younger students especially love to read aloud so that they can practice reading more smoothly and with intonation. This is why partnered reading, buddy reading, or shared reading are popular forms of collaborative reading in the early grades. Working in partners, students take turns reading a portion of the story, sometimes making predictions, or sometimes to simply enjoy a story together.

EXPERIENCES AFTER READING

After-reading experiences help students deepen and extend their understanding of texts. Students can re-live parts of the text through retelling, writing, drawing, dramatizations, and discussions. All these experiences can be done collaboratively, with small groups or pairs of students. Some activities are more appropriate for groups of two, like dialogue journals, while other activities, like skits, are better designed for larger groups of four to six. The following is a list of after-reading experiences which involve peer collaboration:

1. Retellings; each group member contributing part of the story.
2. Retellings; each group member taking a different viewpoint.
3. Skits; groups perform a scene from the story with movement and props.
4. Reader's theatre; group members divide the text and read portions aloud.
5. Pantomime; silent performance using movement and facial expressions.
6. Readers response discussion; groups share personal responses.
7. Video project; group members enact scenes on videotape.
8. Diorama; groups create a particular setting from the book.
9. Dialogue journals; students can write as themselves or as characters.
10. Concept maps; groups create and complete giant concept maps showing literary elements, characterizations, mood, theme, and event timeline.

These collaborative after-reading experiences help students consolidate their understanding of texts, while prompting positive associations between reading and other subjects. Teachers do not need to rely on comprehension questions or book reports to assess student's comprehension of reading material. These traditional closing activities are primarily for individual seat work and are often uninspiring. Providing a choice of various collaborative projects for after reading can get students talking and moving together. Not only are these stimulating ways for students to put closure on a book, but they can also provide teachers with important information about how students transact with others about print.

USING BIG BOOKS

Children can learn skills within the context of good literature with large print and big colorful illustrations. Big books can help connect the positive effects of group reading experiences by letting children observe someone (the teacher) engaged in a desirable activity that they can emulate. Teachers may make the big book available for group readings and small books for the children to use independently. Students might be encouraged to take the small-book version home to share with their family. Activities may be suggested that help children use their interest in the science or math topic covered to move directly into other books or themes. For example, *Pumpkin, Pumpkin* (Tinkerton, n.d.) could lead to a unit on plants or how living things grow and change. Related big books include *The Carrot Seed* (Kraus, n.d.), *The Reason for a Flower* (Heller, n.d.), and *The Little Red Hen* (McQueen, n.d.). A theme on the growth of living things could connect to other big books—*The Very Hungry Caterpillar* (Carle, n.d.), *Animals Born Alive and*

Well (Heller, n.d.), *Little Salt Lick and the Sun King* (Armstrong, n.d.), *The New Baby Calf* (Newlin, 1986), and *Chickens Aren't the Only Ones* (Heller, n.d.).

Many big books will fit into a teachers current theme plans and suggest new topics to explore. Whole classrooms can share memorable stories and interesting facts, and small groups can use big books to encourage listening and speaking. The teacher models reading and risk-free participation by teaching science and math skills within a literary context. Such a meaningful teaching of skills within context is a powerful learning experience. An abundance of good literature about science and mathematics is a key to independent reading.

A big book example in science is *Pumpkin, Pumpkin* by Jeanne Tinkerton. This book has detailed illustrations that blend science with a simple informative storyline exploring a child's wonder at watching a seed grow. It is for younger readers who may need to hear the story several times to build confidence and link the book to common human experiences. A big book example in mathematics is *Bunches and Bunches of Bunnies* by Louise Mathews. This story helps children understand multiplication with a story of rhymes that allow for counting, dramatization, and rapidly multiplying rabbits. Teachers often like to read the story through once and go back to look at the changing rhyme pattern while organizing what is happening on a large chart (the book goes through the numbers one to twelve as the bunnies multiply). Children may be encouraged to creatively strike out on their own with rhyming-couplet patterns that predict the answers to other multiplication questions. *The Silly Song Book* (Nelson, 1981) would connect some of these concepts to music.

On occasion, students might build their own big book in a small group, laminate it, and show it off in class. This way the science and math and reading program is extended to include writing and art. The same possibility exists when using big books like *How Much Is a Million?* (Schwartz, 1898) or *Ten, Nine, Eight . . .* (Bang, 1991).

Using big books for interdisciplinary inquiry in science and mathematics helps accommodate different learning styles by encouraging children to use many and varied resources. In *Ten Little Ducks*, for example, children could take the lead from one of the little ducks and go outside to blow some "magnificent bubbles." This encourages children to piece together meaningful patterns from many sources: literature, science, mathematics, their friends, and the world outside of school.

BOOKMARKS: A TEACHING STRATEGY

Active readers respond to text as they read by monitoring comprehension, predicting, asking questions, recalling related experiences, and self-monitoring comprehension. Good readers also connect life with text and make connections across texts. Strategies that encourage the reader to interact with both the text and peers can result in higher levels of understanding and increased reader involvement. Bookmarks is a simple technique for quickly capturing reactions to text on small pieces of paper as students work to understand a text. Readers can use their brief bookmarks reactions to note their struggles with the unfamiliar and their transactions with vocabulary and concepts. Students can continue to read if they know that they will have a chance to talk about their problems and interpretations after reading:

1. Give students opportunities to talk about their responses to the concepts described in the text before the strategy is introduced. It is often helpful for the teacher to join in the discussions as a reader who has personal responses to the material.
2. Cut construction paper or notebook paper into strips about two inches wide.
3. Teachers do a better job with teaching content areas if they actually read some books and and explore issues themselves. To model the strategy, the teacher can demonstrate by using bookmarks while doing their own reading. The teacher can then share their own responses and demonstrate the process themselves.
4. Students can be given about a dozen bookmarks and asked to choose a piece of literature to read. Ask them to note the page number when they are responding. By having an uninterrupted reading time, it is possible to generate a fair number of responses; these can be stapled together. If you ask the children to always have two books they like (in their desk) it helps with classroom management.
5. Children can share examples of their responses with a small group and selected examples can even be shared with the whole class.
6. The teacher can read through the bookmarks and respond to questions, share reactions, and encourage rethinking or rereading. Some of the bookmarks can be included in portfolios and brief teacher responses can be included on the bookmarks and handed back to the reader (do not grade).

SQ3R

The SQ3R strategy can be used on a regular basis to help students get ready to read (pre-reading). The basic process has five steps: survey, question, read, recite, and review. Students survey the passage to be read by headings, italicized words, highlighted items, questions, and introductory and concluding paragraphs. Next, students come up with questions based on the headings and read the passage to find the answer. In the final review step, students look back over the text and any notes they jotted down while reading to determine the author's major points.

LANGUAGE EXPERIENCE APPROACH

The language experience approach (LEA) is often used in the lower grades to integrate children's knowledge and experience with interdisciplinary inquiry in reading, writing, social studies, the arts, science, and mathematics. This method is based on the notion that what children experience (e.g., a visit to the science museum), they can write about or have someone write for them. What is written can then be read by everyone. With first graders, for example, teachers can print student-generated ideas on a large chart, have the children do some painting, laminate, put some rings through the top, and presto the class has a LEA big book.

RECIPROCAL TEACHING

Reciprocal teaching works well from the middle grades on up and requires the use of four strategies in summarizing content-area material. First, the student reads the passage and summarizes it in one sentence. The next step is to ask one or two higher-level questions about the material read. In the third step, the student explains or paraphrases the most difficult parts of the passage. The final step involves predicting what will occur in the next part of the text to be read. Reciprocal teaching focuses on student dialogue and collaboration in bringing meaning to what is being read in the content area.

A variation of reciprocal teaching is the generating reciprocal inferences procedure (GRIP). Here, students are taught to look for clue words for inference types like location, time, category, action, object, attitudes, instrument, agent, solutions, cause, and effect. Though some of the work must be done alone, students can be divided into pairs so that they have someone to respond to the inferences made from cue words or reading selections. In the beginning, it is often best for the teacher to model the process to show how to find cue words and justify inferences. Two students can also coauthor their own inference paragraphs (including cue words) and exchange their work with other partnerships. The authors can later check to see if the inferences are correct.

READING ON THE GO

This technique is related to the kind of content-area reading that people do (outside the classroom) when quickly reading everyday things like directions for taking medicine, reading the thermometer, understanding highway mileage signs, mixing paint, figuring out a sports column, or calculating the savings in a sale. This kind of reading requires quick judgments. As they work through some real problems in their community or from newspapers and magazines, students can confirm assumptions or note mistakes. This allows children to develop content reading strategies that apply in the world outside of school.

SKETCH A CONCEPT

Whether it is science or literature, we have found that comprehension is helped by having students keep a sketch notebook of illustrations—much like keeping a writing journal. Concepts or story elements can take on visual personas that help with comprehension.

The visual arts can connect to reading passages about science, literature, or stories written by classmates. Einstein, for example, used a similar technique to sketch out the theory of relativity when he was a young man working in Bern, Switzerland. The idea is to use colors, lines, symbols, and shapes to convey meaning. This can be done with a partner or in a small group that focuses on reframing their understanding of the text by sketching a visual representation of the concept. Teachers can set up a whole-class sharing session that integrates the language arts as students think through concepts, listen, speak, "write," and "read" in a visual mode.

USING THE MEDIA TO ENHANCE LANGUAGE LEARNING

Teachers can encourage student involvement in the real world and literature by supplementing text with related topics in magazines, newspapers, and television (Adams & Hamm, 1989). The *New York Times*, for example, has a special science section on Tuesdays. Every newspaper has at least some material that relates to math and science. Even the lower grades can do a little of this by making use of the pictures and weekly-reader-type newspapers that bring things down to their level. The average local paper is written at the fourth or fifth grade reading level and with a little background upper-grade students can comprehend the evening news. This kind of reading, viewing, and discussion can help students above the third-grade level keep up with current events that bypass out-of-date or boring information in some textbooks.

Newspaper Scavenger Hunt

This activity is a good way for students to get to know how a newspaper is organized. Upper-grade students can compare doing the activity using a local newspaper with using a national paper like the *New York Times*. Newspaper Scavenger Hunt can be applied to a variety of reading levels (younger students should use newspapers with a lot of pictures). Pick a day of the week or a newspaper that has a high content of things that you would like your students to know more about (many papers have special days for science, the arts, literature, or whatever). Before class, make a list in two columns of words and phrases extracted from a sample newspaper; you could list cartoons, pictures, graphs, words, concepts, and short phrases. Make enough copies and give one to each pair of students along with the paper. Be sure to have enough newspapers (they must all be the same paper) and give one complete paper to each partnership.

Students work with their partner to find the page where the item is located, put the page number on the answer sheet, and circle the item in the newspaper. A time limit is set for the search to take place; you do not want more than one or two groups to finish. When time is up, the students can compare their "success" rates and check some controversial possibilities. The teacher may focus the discussion on several areas of interest in mathematics and science. This exercise can be modified for a range of ability levels and is a good way for students to get to know the parts of a newspaper and to get ready to actually read some stories in the news.

Reading the Newspaper

After reading a section or article in the print media, students can work in pairs to write a critical analysis using the following questions as guides:

1. Describe the author's basic point in one short paragraph. Note the date, title, publication, page number, and author's name (if given).
2. What have you learned that is unique?
3. What new questions might you raise after reading the article?
4. How do graphs, statistics, illustrations, or pictures back up the main points?
5. Explain one major concept in the passage and point out its connection to math or science.
6. How does the passage you read change your attitude toward the subject and how would you like to solve some of the problems discussed?

Writing a Fictional Story Using Headings and Subheadings

In this activity, students work in pairs or small groups to write a story based on the vocabulary, headings, and subheadings they find in the newspaper. Encouraging humor and wild connections works well here. When finished, the coauthored stories may be read to the group or to the whole class. An example of one line in a news headline and subheading story by middle school students might be, "World Ends: Women and Minorities Hardest Hit."

As they come into contact with what is going on in the world around them, students should be encouraged to unlock the mysteries lying behind the daily headlines. It is much more than language arts. Gaining a familiarity with current events through collaborative experiences with the news media breeds curiosity, enthusiasm, and critical thinking. By improving world knowledge, understanding in all of the basic subjects is amplified.

READER'S THEATRE CAN CONNECT SCIENCE, LANGUAGE, AND LITERACY

Reader's theatre is the oral presentation of prose or poetry by two or more readers. Complete scripts can be provided or students can write them after reading a story or a poem. The actual story or chapter may be ten or twenty pages long, the finished reader's theatre script may only be one, two, or three pages of print. If pictures are used, there may be ten or twelve large illustrations. We recommend trying some prepared scripts first (so that children get the basic idea) and then having the students work in small groups to transform a story or poem into a script.

The typical reader's theatre lesson involves script writing, rehearsal, performance, and follow-up commentary for revision. Before the class presentation, children need a chance to practice and refine their interpretation. Everybody eventually gets their own copy so that they can read their role from a hand-held script (a few mistakes in reading are good for a laugh).

When reading, students stand up or turn to face the audience; when their turn is over, they sit down or turn their back to the audience. If there are four roles and five children, two read the same thing at the same time; if there are four students and six roles, two members of the group read two roles. Lines, gesture, intonation, and movement are worked out in advance. Individual interpretations are negotiated between group members. The performance in front of an audience can intensify the experience and connect the reader to the audience.

Reader's theatre can be a good informal cooperative learning activity, where students not only respond to each each other as character to character but in spontaneous responses that ties the group together with the situation of the text. The idea is to use a highly motivating technique to engage children in a whole range of language activities.

A good script for teaching animals, colors, and sight concepts to first graders is *Brown Bear, Brown Bear, What Do You See?* by Bill Martin (1983). A somewhat similar book (emphasizing sounds) by the same author is *Polar Bear, Polar Bear, What Do You Hear?* Remember that lines and pictures may be changed and shortened by the students to make it work in their reader's theatre. Each child gets two large pictures and words that have been painted, sketched, or copied (enlarged) from the original book. The group gets a chance to practice in the small group and get ready to take their turn in front of the class. When a student's turn comes, they turn to the class and read the words on their picture. The following is a sample script:

1st child shows the Polar Bear
Everyone reads the first line:
> Polar Bear, Polar bear, what do you hear?

1st child reads the next line:
> I hear a lion roaring in my ear (lion turns around).

Everyone reads the next line:
> Lion, lion, what do you hear?

The 2nd child reads the next line, describing what the lion says. This continues until the class gets to the zoo keeper.

Everyone reads the last line:
> Zoo keeper, Zoo keeper, what do you hear?

The last child responds:
> I hear children (pictures put quickly onto the floor in order).

All five children pointing at each picture in turn: I hear children: growling like a polar bear, roaring like a lion, snorting like a hippopotamus, fluting like a flamingo, braying like a zebra, hissing like a boa constrictor, trumpeting like an elephant, snarling like a leopard, yelping like a peacock, bellowing like a walrus; that's what I hear.

THE WRITING PROCESS

Since the mid- to late-1980s, the process writing model has had a major impact on language-arts instruction. Teachers who have adopted this model help their students see writing as a process by having them experience various dimensions of the writing process: inventing, drafting, revising, editing, and publishing. Each of these processes is recursive in nature in that students often revise as they draft or edit as they invent. In fact, problems can occur when teachers see the writing process as a series of rigid steps to be followed in a prescribed sequence (Labbo, Hoffman, & Roser, 1995). The ultimate goal of process writing is to guide students toward developing meaningful written pieces over time.

To write meaningfully, students are encouraged to choose topics that are relevant to their own lives. Consistent with the whole language movement, functional and aesthetic aspects of writing are emphasized. Students write to inform, entertain, persuade, reflect, criticize, summarize, and so on. Depending on the purpose for writing, students are encouraged to experiment with a variety of written genres: expository, fiction, or poetry. Regardless of the genre used, students are encouraged to draw from their own experiences to write.

For the teacher, process writing means writing along with students, modeling not only enthusiasm for writing, but the thinking and composing processes that are an integral part of what real authors do. Teachers share rich literature with children and discuss the literary styles of a variety of authors. They also demonstrate the use of conventions through "mini-lessons," which are tailored to small groups of children who share particular writing needs.

Woven throughout the process writing cycle are opportunities to share writing with others. Talking about topics, drafts, or finished pieces with peers allows students to gain a sense of audience as they think about developing their written pieces. Teachers can stage many collaborative experiences for students across any dimension of the writing process.

Pre-Writing

Inventing texts is a social process and our ideas for composing texts often come from our experiences with others. That is why group sharing is a particularly important part of the rehearsal process. If teachers can stage opportunities for students to first talk about their ideas for composing texts with caring others, students will have a stronger foundation from which to begin composing.

Groups of students can generate ideas for topics in a small group or with a partner. For instance, grade school children often talk to each other about the kinds of stories they would like to write based on their favorite books, super-hero characters, movies, or personal experiences. Calkins (1994) suggests that students use a writer's notebook to jot down the things they notice in and out of school, their memories and ideas, their favorite words and responses to reading, or their conversations with other students. When these notes are shared, peers can then ask questions about these experiences and thoughts, eliciting more information. Such talking opportunities prior to writing can provide authors with a frame from which to begin drafting.

Drafting is a first attempt at writing ideas down in a narrative form. Many teachers invite young authors to turn off their "internal editors" and just allow all of their ideas to flow onto the paper. Once an author is finished with a first draft, they can begin the sharing process which is so essential to refining written pieces.

Peer Revision and Editing

Revision involves making changes in the content of a draft. As stated, revision happens as students are thinking of new ideas to write before they draft, or as they are drafting a piece. Writers tend to revise naturally, without prompting, and at any time during the composing process. Teachers can see this happen as students erase and cross out information as they draft a story. To get students to revisit a piece of writing after they have drafted requires a lot of teacher modeling and peer feedback. This kind of revision should only happen once students are more experienced in drafting (second grade and beyond). It is not recommended for younger students who are just learning how to form letters and write words. The focus of instruction at this level should be to develop interest and purpose for writing.

Peer revision requires comprehension, reasoning, and reflection. Peers can provide a sense of audience to a fellow writer who wishes to make a piece of writing clearer or more vivid. One kind of format for peer revision is the Authors' Circle. While working with three or four students, peers help each other clarify what they want to say. Students are encouraged to bring their pieces to the Authors' Circle when they have come to a point in their writing where they need audience feedback. The process is as follows:

1. The author reads a piece aloud. When writers are more experienced, they can state what they like about the piece, identify problems, and ask for specific kinds of feedback.
2. The listeners tell the author what they heard in the piece and what they found most effective.
3. Listeners raise questions about parts that were unclear or confusing, or parts that need more information. The author can take notes to remember peer comments.
4. Authors can consider the suggestions made by the peer group.

Through modeling, teachers can demonstrate ways to participate in Authors' Circles so that students can learn how to listen and give helpful suggestions to their peers. Once students understand how to give positive feedback on the meaning of pieces, they can then run their own revision groups independently. Students should be able to give feedback on the extent to which the following are true:

1. The piece is focused around one central idea.
2. Ideas are supported by details.
3. One part flows smoothly to the next part (organization).
4. The author's voice is present.
5. Words and phrases are lively and descriptive.

Feedback like this helps students become more aware of how others receive their pieces. During circle time, students can make notes and place marks in places where a part of their writing needs further refinement. Authors' Circles should be combined with teacher–student conferences so that students can get both peer and adult feedback to enhance their written pieces.

As students work to refine the language in their pieces, they also need to consider the clarity and meaning of their message through the use of conventions. Donald Graves (1994) uses the metaphor of sign posts to illustrate the relationship between conventions and meaning: "As I sit at my computer keyboarding these words, every letter that follows every other letter, the spaces in between groups of letters to indicate words, the capital letter at the beginning of each sentence, the period or stop to end an idea, the spaces between lines, all of these are acts of conventions. Like sign posts, they help you, the reader, enter familiar ground so that you can concentrate on the information without distraction" (p. 56).

Teachers need to help students understand the relevance of conventions for conveying meaning. While authors can look at their own pieces to assess spelling, punctuation, or capitalization, sometimes a fresh perspective is needed to help students read their pieces with conventions in mind. Peers can give each other feedback on the conventions of written language through the process of peer editing. Working in pairs, students can swap drafts and give each other feedback on their use of certain conventions.

Another format for peer collaboration would be to set up an editor's desk as a work station in the classroom. Students who enjoy helping others use conventions can take turns sitting at the desk during writing workshop and assisting other students who need help with conventions. This part of the room can include several reference books and wall charts that explain certain conventions.

Peer collaboration can also be used for more explicit teaching of conventions. There is a collaborative game which helps students become familiar with the use of conventions. Working in heterogeneous groups of three, children refer to their own work and a book they are reading to locate specific conventions. Using the following game format, children help each other understand why certain conventions are used:

1. The teacher writes a sentence on the board that shows a particular convention and underlines it.
2. The teacher asks students to look for the convention, first in their work folders, then in a book they are reading.
3. When one team member finds the convention, the other members look at it and discuss its purpose using the following questions: How does this convention help the meaning of the sentence? How does this help readers?
4. When everyone in the group thinks they know how the convention helps the meaning of the sentence, all three group members put up their hands.
5. These group members are asked to explain what the convention is and how it helps readers to understand the sentence better.

Graves (1994) plays this game with students for ten to fifteen minutes every eight to ten school days. The advantage of this game is that students help each other learn about conventions using their own work. Students act as teachers and use relevant materials for instruction.

The success of peer collaboration in writing groups will depend on the way teachers model the use of conventions during mini-lessons. Clear, meaningful modeling using real work in progress will help students understand the use of conventions so that they can draw from the teacher's examples to assist peers. Yet modeling different ways to write and revise is only one consideration for teachers. Teachers need to also understand the ways students feel about themselves and their writing if they want to support students' efforts to help each other.

To a large extent, the view that developing writers come to have of themselves as students and writers is reflected from the response they get to their work. Elementary school children attend carefully to how teachers react to them, and they compare their progress with those around them. They can count stars and smiley faces on their written work or the absence of them. Comparison with peers affects the way students view themselves as readers and writers, and many adolescent writers remember being humiliated in their very early attempts at reading or writing, and these responses caused later struggles in writing.

Before peers can work with each other to revise and edit their written pieces, teachers need to work on establishing trust between students. This involves permitting students some say in the way they are grouped and in the audiences to which their work is exposed. Many teachers permit students to select groups using preset criteria. Some teachers simply say, "Form your own groups of three, making sure that there is a male and a female in each group." You could ask students to find someone with similar shoes for a group of two or ask students to write down three peers with whom they would like to work (the teacher makes the final choice). Try to make the groups small (two, three, or four). A diversity of opinion is important, so you might simply say, "Avoid your friends." If a problem group surfaces, the teacher must work with the group to raise trust or change membership to solve the problem.

Teachers can enhance environments for writing by organizing classroom spaces with quiet isolated corners where students can discuss their work. Supporting each other in observation, critical thinking skills, and the art of writing can help lead to literacy in the fullest sense. Designing writing curricula so that students are writing for peer audiences that they trust, utilizing the workshop model, and designing writing tasks so that peers can work collaboratively toward a common goal foster intrinsic motivation in writing at all levels.

THE WRITING PROCESS APPROACH

In the process approach to teaching writing, teachers use a workshop format where students may bring in writing that they have produced elsewhere. Students can collaborate or work alone on short stories, poems, television scripts, or oral histories with elders. The focus can be on newspaper-generated issues, science fiction, personal experience, or anything else. In the workshop, children talk about their writing, get suggestions, develop new ideas, and are made to feel part of a community of supportive writers. As writers and responders, students are encouraged by the teacher to say something positive, suggest titles, comment on the lead sentence, or ask about the author's favorite part of the story and how it could be made clearer. Students engage in pre-writing activities, write, rewrite, and help each other with editing. After proofreading, students can decide what to "publish." Illustrations may be added, the "book" bound, the cover decorated, a card made up for the classroom card catalog, and the book is placed in the class library to be checked out and read by others.

The development of a writing community is a very powerful way for students to collaborate in developing their writing voice. Whether writing is self, peer, or teacher evaluated, it is important not to lose sight of the connection between what is valued and what is valuable. Jointly developed math and science writing folders (portfolios) have a major role to play in student assessment. By selection of samples, these folders can provide a running record of students' interests and what they can and cannot do.

To work toward less control, teachers need to help students take more responsibility for their own learning. The ability to evaluate does not come easily at first, and peer writing groups will need teacher-developed strategies to help them process what they have learned. The ability to reflect on being a member of a peer writing team is a form of metacognition (learning to think about thinking). The skills of productive group work may have to be made explicit. This requires processing in a circular or U-shaped group where all students can see each other. Questions for evaluative social processing might include the following:

- How did group leadership evolve?
- Was it easy to get started?
- How did you feel if one of your ideas was left out?
- What did you do if most members of your group thought that you should write something differently?
- How did you rewrite?
- Did your paper say what you wanted it to?
- What kind of a setting do you like for writing?
- How can you arrange yourself in the classroom to make the writing process better?
- What writing tools did you use?
- How do you feel when you write?
- Explain the reasoning behind what you did.

Remember, its just as important for students to write down their reasoning as it is to explain their feelings, content understanding, preferences, and solutions to problems.

The recognition of developmental stages in social skills must be taken into account as teachers incorporate literature-based writing concerns into their classroom routines. For younger students, the writing process can take the form of jointly produced language-experience stories that make connections between various elements of the language arts as they are found in the real world. This real-world connection is important in helping students to value reading, writing, listening, and communicating as a way to understand that world and as tools for solving problems. As children work together with concrete materials, they can translate them into stories, pictures, diagrams, graphs, charts, and other symbolic forms. Some of these can be placed on large charts, with the teacher or an upper-grade student doing the writing. As soon as they can write on their own, the children can keep a private journal of drawings, experiences, and writing samples.

As students learn to expand their perspectives, they can begin to carry a story from one page (or day) to the next. Time may be set aside each day for a personal journal entry. Though it is important that the language be in a student's own words, the teacher can make comments without formal grading.

EXPERIENCES WITH POETRY

To read or write poetry involves an awareness of certain elements that make it unique. Teachers must have some basic knowledge of the vocabulary of poetry in order to help children enjoy and mature in their understanding and appreciation of it. Some characteristics include the following:

1. Poetry uses condensed language; every word is important.
2. Poetry uses figurative language (e.g., metaphor, simile, personification, irony).
3. The language of poetry is rhythmical (regular, irregular, metered).
4. Some words may be rhymed (internal, end of line, run-over).
5. Poetry uses the language of sounds (alliteration, assonance, repetition).
6. The units of organization are line arrangements in stanzas or idea arrangements in story, balance, contrast, build-up, surprise, and others.
7. Poetry uses the language of imagery (sense perceptions reproduced in the mind) (Denman, 1989).

Before introducing children to poetry experiences, teachers need to expose children to a wide variety of poems and poets. Many children prefer poems that are humorous, with clear-cut rhyme and rhythm. The narrative is the most popular poetic form and favorite topics are familiar experiences and animals. When reading poetry to children, teachers may have students respond to the images and meanings evoked by sharing their personal interpretations with a partner, similar to the reader-response discussions suggested earlier in this chapter. Students gather in small groups once a week to share poetry books they have been reading. The groups are structured so that each student does the following tasks:

1. Reads the author and title of each book.
2. Tells about the book.
3. Reads one or two pages aloud.
4. Receives responses from members of the group, specifically pointing out parts they liked and asking questions (Nell, 1989).

Poems for older children and adolescents often include figurative language or abstract themes which can be difficult for students to decipher. Sharing interpretations is an excellent way for them to make important connections with poetry.

Poems are also a wonderful medium for helping students develop fluency and confidence in reading. Teachers can model ways of reading poems aloud by experimenting with voice, pitch, and intonation. They can invite students to read their favorite poems to each other by trying out various ways of reading modeled by the teacher. Students sign up and read aloud at the end of each day. Other students "point," commenting on parts of the poem that catch their attention. A classroom anthology of poetry can be illustrated and laminated.

Through a rich exposure to poetry, children will often want to write their own poetry (Sudol & Sudol, 1995). There are many types of poetry students can write that can be enhanced through peer discussions. Students can share their ideas about composing fixed forms of poetry like limericks, haikus, couplets, or free-verse poems that can be humorous, serious, nonsensical, sentimental, dramatic, or didactic. Teachers can also structure collaborative poetry-writing sessions as described in the following examples.

1. Wish Poems

    Each student writes a wish on a strip of paper. The wishes are read together as a whole for the group. Students then write individual wish poems which are shared.

2. Timed Poems

    Divide the class up into small groups or teams. Each team is given a short time (one or two minutes) to compose the first line of a poem. On a signal from the teacher, each team passes their paper to the next group and receives one from another. The group reads the line that the preceding team has written and adds a second line. The signal is given and the papers rotate again—each time the group reads and adds another line. Teams are encouraged to write what comes to mind, even if it is only their name. They must write something in the time allotted. After eight or ten lines, the papers are returned to their original team. Groups can add a line if they choose, and revise and edit the poem they started. The poems can then be read, with team members alternating reading the lines. Later some of them can be turned into optic poems (creating a picture with computer graphics using the words of the poem) or acted out using ribbons or penlights (while someone else reads the poem).

3. Optic Poetry

    Students can make a picture with their poem, repeating lines if needed to fill in the picture. They can be colored in or painted on larger pieces of colored paper and put up around the room. The teacher may ask students to focus on a science or math concept; coauthoring is encouraged.

4. Using Art Reproductions for Poetry

    Teachers or older students may cut up some art magazines. We often use *Art in America* and *Art Forum* and only bring to school those pictures that we would not mind our children seeing. The cost of excellent color copying has really come down and the reproductions can be laminated at school. Each student gets to choose a picture and working alone comes up with

1. One word that describes what they think is happening in the picture.
2. Two adjectives (while looking at the picture).
3. Three verbs ending in ing.
4. A four-word phrase that says something about the picture.
5. One word to sum up. ("What is it?" some students might ask. Finish before saying it is a cinquain poem.)

The second half of the activity involves having students in small groups share their pictures and interpretations. Each group sends a member up to choose a large picture and the group places it on a large piece of poster board. Each student gets a large piece of construction paper and uses a felt pen to write their own cinquain about the one group picture (these should be large enough to be seen from across the room). The whole group construction is put up as public art (with elements of language arts).

An alternative format might be the following:

One word, giving the title.

Two words, describing the title.

Three words, expressing an action.

Four words, expressing a feeling.

One word, a synonym for the title.

5. Poetry with Movement and Music

Poems can be put to music and movement. One student can read the poem and the rest of the group moves around using streamers or penlights in a darkened room.

6. Summarizing with Biopoems (suggested by Jane Kinkle)

Biopoems help students make inferences, synthesize information, and assess the results, all helpful skills in reading and scientific inquiry. Precise language must be used to fit the form and the character. Biopoems have proved to be especially useful in literature and science. They can describe characters in books, plants, animals, planets, or they can be called on to activate a character before writing a story. Biopoems can also be used to help students to get to know each other.

Line 1 First Name

Line 2 Four traits which describe the character

Line 3 Relative (brother, sister, etc.) of

Line 4 Likes (three things or people)

Line 5 Who feels (three items)

Line 6 Who needs (three items)

Line 7 Who fears (three items)

Line 8 Who gives (three items)

Line 9 Who would like to see (three items)

Line 10 Resident of

Line 11 Last name

7. A Lesson Developed from *Dinosaurs* (a poetry anthology for children edited by Lee Bennett Hopkins)

   1. Teacher reads poems aloud.
   2. Students brainstorm reasons why the dinosaurs died, and words that relate to how the dinosaurs moved.
   3. Models of dinosaurs and pictures are displayed and talked about.
   4. Students write poems and share them. Children may also wish to include personal school or family history in their poem. Math and social studies could be incorporated by using population or the area size of a state.

In New Jersey, teacher Jane Kinkle uses poems to help her students connect science to the language arts. For example, "The Microscope" by Maxine Kumin, "The Noiseless Patient Spider" by Walt Whitman, "The Torchbearers" by Alfred Noyes, and "Space" by Anne Morrow Lindbergh.

LANGUAGE, LITERACY, AND CREATIVE DRAMA

Creative drama has many facets and applications in the language-arts classroom. It emphasizes spontaneous, intuitive, and natural responses to literature in the classroom as opposed to prepared or staged theatrical productions that require much advance preparation (Stewig & Buege, 1994). Creative drama allows students to bring their own worldviews and experiences to texts by using movement, facial expressions, or language through interpretive or improvisational activity. Interpretive experiences allow students to evoke their own renditions of characters or events that bear a relationship to a given story. Improvisational experiences allow students to explore characterization or plot by adding new conditions or considerations to stories, thereby reinventing stories in creative ways. Creative drama does the following:

- Adapts to many types of lessons and subjects.
- Encourages the clarification of values.
- Evokes contributions and responses from students who rarely participate in "standard" discussions.
- Helps in assessing how well students know the material (characterization, setting, plot, conflicts).

- Provides a stimulating pre-writing exercise.

Drama is a natural way to improve vocabulary, listening, observation, and speaking ability. Skills learned through creative dramatics can carry over into the writing process, in part because drama engages students in the rough equivalent of pre-writing, writing, conferencing, and rewriting. Creative drama can also help motivate students to clarify concepts, pay more attention in reading, and explore the deeper meaning in the text. Within these activities, students can be viewed as performers required to demonstrate their collective knowledge. The teacher's role is like that of a coach—helping students know and interpret the standards provoking thought. The following is a list of ways creative dramatics can be used.

1. Stimulating Exploration of the Environment

    A stimulating environment for language learning includes tools for observation like microscopes, maps, thermometers, and telescopes. These materials stimulate the reasoning, explanation, comparison, drama, and language of observation. Teachers tend to think of these materials and methods as parts of the science or social studies curriculum. They are also crucial to language learning.

2. Personification (this can also be used as a pre-writing activity)

    Each student draws the name of an inanimate object (pencil sharpener, doorknob, waste basket, alarm clock, etc.). Students pick a partner and develop an improvisation from the objects in a story or in the environment.

3. Increasing Research and Journalism Skills

    Using techniques of role playing and creative drama, have student groups show how to interview (give good and bad examples). Short excerpts from TV news or radio information programs provide good models for discussion and creative drama activities.

4. Showing Emotions (non-verbal communication)

    Assign an emotion to each student (anger, jealousy, shyness, nervousness, arrogance, etc.). Students must act out the emotion without actually naming or referring to it. The class notes significant details and discusses which emotion was being portrayed.

5. Extend a Story

    Let students speculate on "what's next" in the life of a character based on what we know from the literature.

    Ask "What happens to _____ next?"

    Ask "What if _____ and _____ met a year later?"

    Ask "What advice would _____ give to someone in a similar situation?"

    Have students act out what they think should go in the blanks, or come up with a new ending to a story they already know.

Students must be given opportunities to engage in exploratory talk with peers. Incorporating cooperative learning structures into the classroom routine is consistent with holistic notions of social learning, student empowerment, and improving all of the language arts. Collaborative learning allows students to see each other as legitimate literacy partners and teachers. Having multiple peer teachers is critical for meeting the diverse needs of developing readers and writers in today's language-arts classroom. Of course, it helps if students can express themselves well. This means that teachers may have to give a little more attention to speaking; a neglected element of the language-arts curriculum.

TEACHING STORY DRAMATIZATION

Try the following exercise in a class:

1. Select a good story, and then tell it to the group.
2. With the class, break the plot down into sequences, or scenes, that can be acted out.
3. Have groups select a scene they wish to dramatize.
4. Instruct the groups to break the scene or scenes into further sequences, and discuss the setting, motivation, characterizations, roles, props, and so on. Encourage students to get involved in the developmental images of the characters—what they did, how they did it, why they did it. Have groups make notes on their discussions.
5. Meet with groups to review and discuss their perceptions. Let them go into conference and plan in more detail for their dramatization.
6. Have the whole class meet back together and watch the productions of each group. Instruct students to write down five things they liked and five things that could be improved in the next playing.
7. Let the players return to their groups at the end of all group performances and evaluate the dramas.
8. Allow groups to bring back their group evaluations to the whole class. Discuss findings, suggestions, and positive group efforts.

By connecting creative drama to the discussion of literature and writing, students can share imaginative ideas with their peers and create an atmosphere where unconscious thought can flow freely.

THE INTEGRATED TAPESTRY OF LANGUAGE AND LITERACY DEVELOPMENT

Whether it is for scientific inquiry or the interpretation of textual material, in reading we learn language to make sense of the world around us and to communicate with others. Learning to read begins even before children come to school. The way we, as adults, learned to speak, read, write, and use communications technology is bound to influence our attitudes about literacy and how we think it should be taught. What do you recall about the process? What could be done better? In our workshops, we have teachers write a short autobiography about how they learned to read. Prospective teachers may have to call their parents and talk it over. This kind of reflection is a good way to link critical thinking to transformative action.

Though it is sometimes useful to consider various aspects of language and literacy separately, it is important to emphasize the complex interactions that take place between them. Learning is more complete when teachers recognize that each dimension of language learning overlaps with others. In much the same way, standards take on more meaning when students are motivated to integrate their knowledge into their lives outside of the classroom.

Language and literacy are essential to mastering what needs to be known to become productive and self-reliant members of a democratic society. Being a critical language user means thinking, questioning, making inferences, and commenting on what is viewed, heard, and read. This involves structuring informed decisions based on prior experience, connections between concepts, and multiple forms of evidence. Engaging in integrated, functional, and meaningful interaction with other people allows children to construct their own knowledge about language.

There is general agreement that a language-rich home and school environment is key to any conception of literacy. As far as teachers are concerned, it is important to be sure that students can read, write, talk, and use appropriate technology across the curriculum. This involves connections to real-world situations, a significant amount of literature, and making sure that students can use spoken, written, and visual language well enough to communicate effectively with a variety of audiences.

Better thinkers, problem solvers, and readers are in control of their own future—and once there they make better strategic decisions based on what they expected to find.
—M. Pressley (1987)

Chapter 7

Interdisciplinary Themes in Science and the Language Arts

> *Some important themes pervade science, language, communication, and technology. They appear over and over again, whether we are looking at an ancient cave painting, a Shakespearean drama, Newton's laws, the human body, a comet, or a book. They are ideas that transcend disciplinary boundaries and influence our thinking and our values.*
> —Rutherford and Ahlgren (1990)

The spirit of inquiry can be strengthened when science, language, and literacy are embedded in thematic lessons. Though some of the themes presented here have language and literacy at the core, the integrated science content of the earth, life, and physical sciences get most of the attention.

It is important to broadly define the content that students need to know in order to become informed, confident, and competent. Overarching big ideas (unifying concepts and processes) of science and the language arts include some of the most important scientific and linguistic concerns. Research supports the use of such integrated inquiry techniques to achieve higher levels of thinking and to make learning more meaningful (Bybee & DeBoer, 1994; NCTE, 1996). Such an approach is viewed as a way of teaching and organizing instruction so that selected elements of subject matter are related to each other.

TEACHING THE BIG IDEAS OF SCIENCE AND THE LANGUAGE ARTS

Unifying concepts are often called the big ideas or themes of a discipline. There are several criteria developed by Post, Humphreys, Ellis, and Buggey (1997) and Martinello and Cook (1994) to help teachers decide if a theme is important and meaningful:

1. Is the big idea relevant, timely, and important?
2. Does it broaden students' understanding of the world?
3. Is the big idea interdisciplinary and does it connect to other knowledge?
4. Does the theme relate to the genuine interests of students?
5. Does interdisciplinary work lend itself to student inquiry?
6. Does the topic include a collection of resources and technology?
7. Does it require planning?
8. Are the investigations designed to be cooperative?

CONCEPTS AND PROCESSES OF ELEMENTARY SCIENCE AND LANGUAGE ARTS

The goal of the National Science Education Standards (NRC, 1995) in meeting the unifying concepts and processes is to provide students with understandings and abilities that align them with the following paraphrased themes from the National Science Education Standards and the Standards for the English Language Arts (NCTE, 1996):

1. Systems, Order, and Interactions

    A system is an organized collection of things that can have some influence on one another and appear to constitute a unified whole (Rutherford & Algren, 1990). Some examples of systems are the number system, education systems, writing systems, the solar system, weather systems, oxygen systems, monetary systems, systems of time, measurement systems, garbage systems, telephone systems, electric systems, sound systems, the communication system, and so on. Think of everything within some boundary as being a system. Students can form an understanding of order in systems. Nature is the same everywhere. There is a sense of regularity where events can be predicted and described. Learners then can develop understandings of basic principles, laws, theories, or models to explain the world.

2. Evidence, Models, and Explanations

    A model of something is a simple representation that can help others understand it better. A model may be a plan, a man-made device, a drawing, or a computer program. For example, a pump has been used as a model to represent the heart, and a pen is a model which represents writing.

3. Constancy, Change, and Measurement

    Constancy refers to ways in which systems do not change (a state of equilibrium). Change is important for understanding what will happen, as well as predicting what will happen.

4. Evolution and Equilibrium

    The idea of evolution is that the present arises from the forms of the past. All natural things and systems change through time.

5. Energy, Form, and Function

    Energy is a central concept of the physical sciences that pervades mathematical, biological, and geological sciences because it underlies any system of interactions. In physical terms, it can be defined as the capacity to do work or the ability to make things move; in chemical terms it provides the basis for reactions between compounds; and in biological terms it provides living systems with the ability to maintain themselves, to grow, and to reproduce.

6. Language and Language Structure

    Language is the most powerful available tool we have for communicating our thoughts, defining our culture, and representing ourselves. Learning the structure of language enables us to use language effectively throughout our lives, whether reading, writing, experimenting, viewing, or listening. The study of the systems and structures of language conventions including grammar, punctuation, and spelling allows students to apply their knowledge, gain experience, and adapt language to different tasks and audiences.

7. Reading

    Reading a wide range of texts and literature is part of science and language learning. This includes classic, contemporary, and popular narratives, poems, songs, lab reports, and plays. Exploring literacy gives students new perspectives on their experiences and allows them to discover how literature can make their lives richer and more meaningful.

8. Researching

    As students engage in research on issues and interests by asking questions, posing problems, and generating ideas concerning language, science, and technology, they accumulate, analyze, and evaluate data from many sources to communicate information and their discoveries in ways that meet their purpose.

Knowledgeable teachers are incorporating thematic strands into their curricula. Themes provide a unifying structure that is helpful in guiding teachers as they develop instructional tools. They can be used to integrate concepts and facts throughout all areas of the curriculum. Some science educators are concerned that thematic scientific inquiry means that the usual curricular divisions of earth, life, and physical sciences may be diminished. The same concerns are voiced by language-arts classroom teachers. Just the opposite is true. As disciplines rapidly expand, a thematic approach serves as a powerful way of uniting or transferring knowledge from one field to the next. If these connections are successful, it is hoped that these intellectual habits will carry over and enrich other fields and disciplines. As a result, students may start to see the overall purpose and logic of the educational system more clearly. Thus, an integrative approach to inquiry will not only help them develop a meaningful structure for understanding science and language arts, but also allow them to see the relationship to other subjects and their daily lives.

DEVELOPING AN INTERDISCIPLINARY THEMATIC UNIT

The interdisciplinary thematic aspect of today's literacy builds on the way citizens actually use language and the scientific method to study the natural world much as scientists do. Wei Wei Cai (1997) suggests using an interdisciplinary approach to building thematic units This way teachers can examine themes by applying methods and language from different disciplines to examine an issue, problem, experience, or central theme. Attacking a broad topic with the intellectual tools of several fields stimulates critical thinking skills and can easily allow for learning through social interaction. Motivation can be intensified when student interest is central to the choice of theme or problem to explore.

A thematic unit is more than a collection of lesson plans. It should be viewed as a dynamic inquiry project. The basic goals are set ahead of time through a joint teacher–student effort. The steps to a thematic unit are as follows:

1. Selecting a Theme

    It should be challenging and related to real-world concerns. By building on the students existing knowledge, an interdisciplinary thematic unit needs to be rich enough to hold interest for at least a week. This means connecting student interests to the curriculum in a manner that allows for the development of interests and academic skills.

2. Deciding on a Desired Outcome

    Unit outcomes need to be decided early on. These may relate to cognitive skills (comprehending concepts), affective development (attitudes), and social development (working in groups).

3. Mapping and Brainstorming

   This is the stage of idea collection and organization. Graphic organizers can be used to outline the major activities for each subject area. As far as brainstorming is concerned, all the possibilities can be listed without criticism, and then the critique can start and the list pared down.

4. Making a Timeline

   At this stage the teacher is the key decision maker. A length of time is allotted for each activity and learning experience.

5. List the Concepts and Skills

   List everything that will be part of the process.

6. Resources

   Outline what materials everybody needs.

7. Learning Centers and Bulletin Boards

   These can serve as vehicles for reaching unit outcomes. An interactive bulletin board can connect to e-mail or even the Internet.

8. Cumulative Activity

   At this point students should be able to synthesize what they have gained from the various disciplinary tools applied to the problem.

9. Assessment Plan

   This may include performance assessments, portfolios, conferences, anecdotes, and exams. At least an informal evaluation is need to make it better the next time around.

10. Daily Lesson Plans

    These need to include a specific description of lesson objectives, rationale, concepts, materials, and procedures. Things can always be changed as you go along.

When thematic units are participatory, rich in content, and related to student interest, they can inspire enthusiasm in both teacher and class.

CONNECTING SCIENCE AND LANGUAGE-ARTS CONTENT

The units and activities presented here are designed to give elementary students a chance to thoroughly explore a science and language-arts concept. It is hoped students will enjoy the activities when they are given a chance to engage in scientific inquiry. Collaborative learning is encouraged, as teachers strive to integrate the disciplines of science and language arts in interesting ways. Whether groups are chosen by the teacher or selected by the students themselves, teamwork adds to almost any lesson.

Each activity focuses students' attention on some interesting event or method. Materials that are needed for the activity are identified and followed by step-by-step directions. Occasionally, some information or method is added within the procedures to help clarify the students' experience. An assessment section is included in most activities. Performance assessment offers opportunities to inquire and learn more about a topic. A background-information section in some of the activities provides needed information or notes to help the teacher. The goal is to provide teachers with background experiences that will aid in building a repertoire of strategies, activities, and skills for teaching science and language-arts content to children.

EARTH SCIENCE

Unit 1: The Study of Rocks and Geology

There are many questions about how life first evolved. One way we can trace the history of the earth is to teach about the processes that explain the origin of rocks and their changes.

How Are Rocks and Minerals Formed?

Rocks are made of minerals, natural inorganic materials that make up much of the earth's crust. Minerals are made from pure elements in the earth that combine to form many different substances. Geologists have developed many ways of identifying minerals. These may include several characteristics, including hardness, color, how it splits, how it breaks, density, and texture. Most rocks contain several different minerals. Some are formed under heat and pressure from inside the earth. Others are deposited in wet and cool environments by water, wind, and ice. Some rocks are melted inside and cooled into rock again. The same processes that formed rocks in the past continue today. Over many thousands of years, rocks change their forms. There is much evidence that the same mineral materials are used over and over in a kind of rock cycle.

Some History of Rock Formation

One of the most violent ways rocks are formed is from the eruption of volcanoes. The rocks are composed of a hot fluid called magma that circulates deep within the earth. Igneous rocks form when the magma cools and solidifies. When a volcano erupts, the force causes some of the magma to spew forth into the outside. This molten material is called lava.

The interaction of the air and hot lava forms an extrusive igneous rock (one that is extruded or forced out of the earth and then forms on the surface). Basalt is a type of extrusive igneous rock. Igneous rocks can also form slowly inside the earth. The intrusive igneous rocks are made from magma that cools and hardens slowly. Granite is a type of intrusive igneous rock.

Sedimentary rocks form on the earth's surface from particles of minerals, other rocks, organic remains, and or chemical precipitates. These rocks are compacted and cemented by the earth's processes. It takes millions of years for those small particles or sediments to become rocks. The layers of sediment press down upon one another, and then they are bonded or stuck together. The bonding agent is a mineral called calcite, found in the water that percolates between the sediment beds. The calcite acts as nature's cement, and causes the grains of sediment to stick together and form sedimentary rock. Sedimentary rocks are the most common rocks, and they are all formed in a similar way. Sandstone and limestone are examples of sedimentary rocks.

Sometimes heat, pressure, gases, and other earth processes act upon igneous and sedimentary rocks in a way that changes the composition of their minerals, their structure, or their texture. The rocks that are produced from this action are called metamorphic. Like igneous and sedimentary rocks, metamorphic rocks may be coarse or fine grained in texture. Some examples of metamorphic rocks are marble, coal, and diamonds.

Sample Activities: Rocks

Activity 1: Finding Out about Rocks

The first activity is an introduction to a unit on rocks. Language arts and writing skills are emphasized. Students have brought a rock to class. They get to know their rock by describing it to someone else. They touch their rock. They then pass their rock to the teacher, who collects the rocks in a box. Then the teacher tells the students he or she will pass the rocks back. The students must get in a circle, put their hands behind their back, and pass the rocks to the person next to them. They cannot look at the rock in their hand. If they think the rock is theirs, they can take a peek. If successful, they take their rock and sit down. This continues until everyone has found their rock. Next, students do guided imagery or the visualization activity. After the visualization, students share what happened in their visualization. Students also write a poem about their rock.

Themes: evidence, models, explanations, systems.

Activity 2: Rock Guided Imagery Experience

Guided imagery is much like a story. The teacher guides students through an imaginary journey, encouraging them to create images or mental pictures and ideas. This activity should be done in a quiet relaxed atmosphere.

Teachers may wish to dim the lights or have students rest at their desks while they read the visualization. After reading, students follow up with some kind of creative activity: discussing their experience in their group, writing in their science log, or creating an artistic expression of some kind. This is a good way to start a thematic unit on rocks. Students write or share ideas about their rock.

Themes: scale and structure, patterns of change.

Guided Fantasy: A Rock

> Close your eyes and imagine that you are walking in a lush green forest along a trail. As you are walking you notice a rock along the trail. Pick up the rock. Now make yourself very, very tiny, so tiny that you become smaller than the rock. Imagine yourself crawling around on the rock. Use your hands and feet to hold onto the rock as you scale up its surface. Feel the rock. Is it rough or smooth? Can you climb it easily? Put your face down on the rock. What do you feel? Smell the rock. What does it smell like? Look around. What does the rock look like? What colors do you see? Is there anything unusual about your rock? Lie on your back on the rock and look at the sky. How do you feel. Talk to the rock. Ask it how it got there, ask how it feels to be a rock. What kind of problems does it have? Is there anything else you want to ask the rock or talk to the rock about? Take a few minutes to talk to the rock and listen to its answers. When you are done talking, thank the rock for allowing you to climb and rest on it. Then carefully climb down off the rock. When you reach the ground, gradually make yourself larger until you are yourself again. When you are ready, come back to the classroom, open your eyes, and share your experience (Hassard, 1990).

In their portfolio, instruct students to make a list of as many observations as they can. Then direct them to write a Japanese poem following these directions:

Line 1—Identify the object.

Line 2—Write an observation of the object.

Line 3—Write a feeling about the object.

Line 4—Write another observation about the object.

Line 5—End with a synonym for the name of the object.

Themes: language, language structure, reading, systems, patterns of change.

Applying the Science Standards: Earth Science Activity

The Cookie/Mining Simulation

This hands-on simulation is intended to introduce students to interdisciplinary concepts and processes which allow students to study real-life problems and problems related to language arts, science, and technology through the use of activities that involve hands-on manipulation. Students should have some knowledge of an operating mine (books, science magazines, and videos are helpful), and should have learned some of the difficulties involved in reclaiming land after mining is completed.

Procedure

1. The class begins with a discussion of what students know about mining. Encourage students to brainstorm and write down all the things they know about mining, both positive and negative.
2. Student participants will imagine that they are miners.
3. Form groups of five miners. Each miner chooses a cookie and a tool for mining. Your group is a mining business.
4. Each miner receives a cookie and a piece of graph paper. Have students trace their cookie on the graph paper. This will be used at the end of the mining excavation.

Rules

1. No player can use their fingers to hold the cookie. The only thing that can touch the cookie are the mining tools and the paper the cookie is sitting on.
2. Players should be allowed a maximum of five minutes to mine their cookie. Your mining group counts the total number of chips discovered.

Evaluation

1. When finished mining, discuss reclamation efforts. Groups compare their mined cookie to what the outline on their graph paper shows. At the end, all players get to eat the remainder of their cookie.

Unit 2: Investigating Water, Air, and Weather

In this section, some characteristics of water and air will be examined, as well as the importance of a having a clean supply of both. The section also presents the following information basic to understanding weather:

1. properties of water
2. air and its properties
3. meteorology and weather

Properties of Water

Water is necessary to life. Though people may survive for weeks without food, water is needed within a few days. Many vital activities performed by humans involve water: bathing, cooking, agriculture, energy resources, and recreational activities, to name a few.

Because water has so many uses, it is important to understand its underlying characteristics:

1. Water dissolves more substances than any other liquid, making it an excellent solvent.
2. Water has molecular attraction called *cohesion*. Cohesion of water molecules is central to the process of evaporation. Water's cohesion allows heat energy to evaporate water more quickly.
3. Water has pressure, caused by its weight. Pressure is involved when something floats.

Despite the fact that 71 percent of earth is made up of water, most is in the form of salt water in the oceans (too salty for plants or animals). Fresh water found in lakes, rivers, reservoirs, and ground water beneath the surface is the immediate source of usable water. The water cycle plays a big role in the availability of fresh water. A steady supply of ground water and other water are supplied by the oceans. When salt water evaporates, the salt is left behind and the wind carries the water vapor far inland. There, it condenses and falls back to the earth.

Activity 1: Exploring Water Cohesion and Surface Tension

How many drops of water will fit on a penny?

Materials

one penny for each student, glass of water, paper towels, eye droppers (one for each pair of students)

Procedure

1. Have students work with a partner.
2. As a class, have them guess how many drops will fit on the penny.
3. Record the guesses on the chalkboard.
4. Ask students if it would make a difference if the penny was heads or tails.
5. Record these guesses on the chalkboard also.
6. Instruct the students and their partners to try out the experiment by using an eye dropper, a penny, and a glass of water.
7. Encourage students to record their findings in their science and math log.
8. Bring the class together again and encourage students to share their findings. Record their responses on the chalkboard.
9. Introduce the concept of cohesion. Cohesion is the attraction of like molecules for each other. In solids, the force is strongest. It is cohesion that holds a solid together. There is also an attraction among water molecules for each other.

Interdisciplinary Themes 139

10. Introduce and discuss the idea of surface tension. The molecules of water on the surface hold together so well that they often keep heavier objects from breaking through. The surface acts as if it has a "skin" on it.

Evaluation
1. Have students explain how this activity showed surface tension. Instruct students to draw what surface tension looked like in their science and math log.
2. What made the water drop break on the surface of the penny? (gravity)
3. What other examples can students think of where water cohesion can be observed? (rain on the car windshield or a window in the classroom, for example)

Themes: systems and interactions, patterns of change.

Air and Its Properties

Air occupies the space all around the earth. Air is found in soil, in water, and in such porous materials as sponges, bricks, wood, and even bread. Air has many of the same properties as water and may contain water in the form of water vapor. The properties of air are outlined here:

1. Air is a mixture of many gases. The two principal gases in air are oxygen (21%) and nitrogen (78%). Together, oxygen and nitrogen make up almost 99 percent of air. The remaining 1 percent is made up of carbon dioxide. The gases that make up air are colorless, odorless, and tasteless.
2. Air has no shape of its own, but assumes the shape of its container.
3. Though air is invisible, it is real and takes up space.
4. Air has weight (1 liter of air weighs about 1 gram).
5. Air has pressure. Air has weight, and anything that has weight pushes or presses against things. As people go higher into the air the pressure becomes less. Moving air (wind) exerts pressure.

The following activities provide some examples of the properties of air (Victor & Kellough, 1994), and illustrate the themes of systems and interactions and patterns of change.

Activity 2: Observe That Air Occupies Space

Background Information
When a group of first graders are asked about air, they seem to know basically what air is, and that it exists. They know that it is everywhere, and that it takes up space, so this is an activity to prove it. Children will have a great time actually "seeing" how air takes up space.

Objectives
Students will understand the concept of air after doing this experiment. They will see how air takes up space.

Skills

observing, inferring

Materials

paper towel, tumbler, aquarium, water

Procedure

1. Instruct students to crumble a dry paper towel and stuff it into a tumbler so that it will not fall out when the tumbler is held upside down.
2. While holding the tumbler upside down, push the tumbler straight down to the bottom of an aquarium or large glass jar that is filled with water.
3. Direct students to observe that the water does not fill the tumbler.
4. Ask students to explain why this is so. (The space in the tumbler is occupied by air.)
5. Tilt the tumbler slightly. Students will be able to see air as it escapes from the tumbler in the form of bubbles.
6. Now lift the tumbler straight out of the water and remove the napkin.
7. Have students note that it is still dry.

Evaluation

1. Have students write their remarks about this experiment in their science and language-arts journal.
2. Did anything surprise them?
3. Have students think of other experiments that would show air takes up space.

Themes: systems and interactions, patterns of change.

Activity 3: Investigate That Air Exerts Pressure

This demonstration attempts to prove that air exerts pressure in all directions.

Materials

tumbler (or plastic glass), cardboard (or cover from a cottage cheese container), water

Procedure

1. Fill a tumbler or plastic glass with water.
2. Place a piece of cardboard or plastic cover on top of the glass and hold it firmly against the glass with the palm of one hand.
3. Grasp the base of the glass with the other hand and quickly turn the glass upside down over the top of a sink.
4. The demonstrator carefully removes the hand from below the cardboard or plastic cover, being careful not to jar the cardboard or the glass. The cardboard and glass will remain in place.

Interdisciplinary Themes 141

5. Ask the students what happened. (The water stays in place because air is exerting a pressure on the cardboard. The pressure of the air against the cardboard is greater than the pressure of the water against the cardboard.)
6. Turn the glass sideways and in many other positions. The water will stay in the glass, showing that air exerts pressure in all directions.

Evaluation
1. Invite the students to try the activity for themselves.
2. Encourage students to respond in their journal to the following questions:
   What happened when you let go of the cover?
   What happened when you tipped the glass in all directions?
   What do you think would happen if you used a half-full glass?

Activity 4: Observe That Air Has Weight

Materials
   two matched balloons, meter stick, scissors, tape, string

Procedure
1. Hang a meter stick evenly from a doorway or other place. Use a string and tape.
2. Attach a string loosely to each of two inflated balloons. Large round balloons the same size work best; blow them up so they are the same size when inflated.
3. Tape each string end to an end of the meter stick. Be sure the stick is level after the balloons are hung. If not, place a partly open paper clip on the stick where needed to balance it.
4. Puncture one balloon with a pin. The deflated balloon will not weigh as much as the balloon that still has air in it, and the meter stick will become unbalanced. (When the balloon bursts, a piece or two may be blown off. Be sure to collect them and drape them around the balloon. Otherwise the results will be inaccurate.)

Evaluation
1. Encourage students to think of other experiments they could try to prove that air has weight.
2. Have students write their reactions to these properties of air activities in their science and language-arts journals.

Themes: systems and interactions, patterns of change.

Collecting Weather Data: Meteorology

Meteorology is the science that studies the atmosphere and weather. For many years, meteorologists have collected data on the earth's surface to predict weather patterns. Their ability to monitor the earth's surface and alert the public of weather dangers has been greatly increased with the new advances in weather satellites. The complex interactions between earth's masses of air, water, and land are being analyzed to yield a further understanding into the kinds of efforts that must be attempted to understand global climate problems (e.g., atmospheric warming and pollution).

A meteorologist is a professional who studies and forecasts the weather. Meteorologists can make fairly accurate weather forecasts by collecting the following information:

1. the temperature.
2. the air pressure.
3. the direction and speed of the wind.
4. the humidity.
5. the kind and amount of precipitation.
6. the condition of the sky.

What Are the Natural Elements of the Earth's Climate and Weather?

People live on the surface of the earth, surrounded by a shield of air called the atmosphere, which is needed to live. Air is needed to breathe. The atmosphere also produces rain, on which life depends. Rain falls from clouds, which consists of tiny particles of water or ice. Clouds are moved by the wind. Sometimes clouds are seen as growing bigger or smaller.

Heat from the sun warms the surface of the earth. Because a large part of the earth's surface is covered by oceans, the sun heats the water near the surface of the oceans, holding it for a long time and releasing it very slowly. The water is continuously evaporated into the atmosphere to form clouds. Clouds may move a long way before producing precipitation. Rain falling on the ground may soak into the soil right away or flow quickly into streams or rivers. Eventually, the water finds its way back to the ocean.

Colder air moving over the oceans is warmed by the water. Warm air is less dense than cold air, and thus the cold air moves in and pushes the warm air upward. The rising air cools and descends again. This repeating cycle is the invisible "engine" that causes the winds to act as they do. Wind is the movement of air set up by the unequal heating of the earth's surface by the sun. Warm winds passing over cold regions manage to warm them. In many places, the wind blows from the same direction most of the time (prevailing winds).

Warming of the atmosphere results in varying amounts of atmospheric pressure, which can be measured by a barometer. Meteorologists depend on many thousands of stations where the atmospheric pressure is measured regularly. The earth's axis is inclined; as a result, the region with the most intense sunlight moves northward and southward over the year. At the same time, the length of the day varies. These effects influence the seasons. When it is summer in the northern hemisphere, it is winter in the southern hemisphere. Climate is the daily and seasonal weather of a region over a period of time.

Water constantly evaporates from the earth's surface. The water is transported by the winds, often over long distances. Water vapor forms clouds that release precipitation as rain, snow, and other forms.

How Are People Affected by Weather?
How Can Weather Be Predicted?

Weather forecasting is difficult and very important, because much of what people do is determined by the weather. Forecasting for more than a few days is very uncertain, but with more and more sophisticated computer technology and better data-gathering networks, meteorologists will soon be able to do a better job of forecasting for a longer time in the future.

Certain types of clouds are strongly associated with types of weather. Because of this, clouds are useful in short-term weather forecasting. Puffy white cumulus clouds are associated with fair weather. Sometimes small morning cumulus clouds grow into huge towering cumulus in the afternoon and produce rain showers or thunderstorms.

LIFE SCIENCE

Children are extremely interested and concerned about all aspects of life, continually expressing their curiosity. They want to know what happened to the pretty yellow and black bird they saw on the bird feeder, they want to know what the name of the butterfly is outside on a leaf, and they want to know about life itself. Young students wonder if rocks are alive, and older students wonder how life itself began. The teaching of life science will help teachers develop and extend children's inquisitive desire to learn, broaden their knowledge, and foster concern for life in its many diverse forms.

Plants, Animals, and Other Living Organisms

Living things have characteristic traits by which they can be described and distinguished from non-living things. Living things take in nutrients and give off wastes. They grow, reproduce, and respond to stimuli from their environments. All living things need certain resources to grow, such as air or other gases to breathe, water, and food. If any of these resources are lacking, the organism (or living thing) will die. Plants and animals are living things.

Life Science Units for Primary Students: Some Suggestions

1. Have students tend to plants and animals in the classroom. Focus students' attention on what a living thing needs for a healthy life, and how human needs are similar or different. Stress responsibility and cooperation.
2. Have students observe animal behavior. Common pets, such as cats and dogs, can be brought by students from their homes to school. Other observations may be of earthworms, goldfish, or hamsters. For these nondomesticated animals, set up work stations with a different animal at each. Allow students to rotate through the stations, observing, taking notes in their journals, and comparing observations.
3. Have students experiment and grow plants. Bean seeds are popular because of their fast growth period, but many seeds can be grown, observed, measured, graphed, and reported on.
4. Encourage students to observe and measure the reactions of plants to growing in light versus dark environments, to different amounts of water, to temperature fluctuations, and so on.
5. Have students classify animals and plants. Classifying leaves is a good example.

Life Science Units for Older Students: Some Suggestions

Some of the same units and activities can be used with older students. The microscope can be introduced and used to enhance observations. Some experiments may include the following:

1. Observing organisms in pond water using a microscope.
2. Looking at cells under a microscope.
3. Controlled experiments with plants.
4. Continued attention to classroom animals and observation of animal behaviors.
5. Growth of corn (or popcorn) plants (Howe & Jones, 1993).

The themes covered by these activities are systems and interactions and patterns of change.

Sample Activities: The Functions of Living Things

Activity 1: Outdoor Field Trip—Plant Characteristics

Take an outdoor field trip within walking distance of your school. Have students look for plant characteristics. Encourage students to take leaf and bark rubbings using crayons and construction paper. Collect weeds, blossoms, twigs, needles, and so on. Then have them compare the differences among plants. Observe and listen to students' comments and conversations. Ask questions about plant characteristics and structures. When arriving back at the classroom, have students classify and label the parts of the plant. Have the students include these observations and charts in their notebook or portfolio.

Activity 2: Explore How Seeds Grow— Creative Movement

This activity involves students in an active spatial form of expression. Themes covered are systems and interactions and patterns of change.

1. Have the class pretend they are a seed. Instruct students to curl up on the floor and pretend they have just been planted in the ground.
2. It is starting to rain. Guide students in their thinking that the seed starts to take in moisture.
3. The seed swells and a tiny plant starts to grow inside the seed. Encourage the students to act out the process.

PHYSICAL SCIENCES

The thematic inquiry skills presented in the earth and life sciences will continue in the physical sciences. What do we mean when we say physical science? Physical science is the branch of science concerned with the nature and behavior of matter, energy, and forces of the universe. This section begins with unifying theories concerning a definition of the structure of matter; questions and sample activities follow.

Exploring the Composition of Matter

Unifying Ideas

Matter is what scientists call the substance that makes up the world and the universe. In the study of matter, the central principal is that the amount of matter stays constant. It cannot be created or destroyed; only its form is changed. From stars to dust, from elephants to ants, everything is made of the same basic building blocks. Though the scale of our universe ranges from the very large to the very small, all of it is matter. The basic premise of the modern theory of matter is that the elements consist of a few different kinds of atoms that join together in different ways to form substances. There are one or more of these atoms for each of the approximately 100 elements.

All matter has properties that can be observed, defined, and recorded. Matter occupies space, it has substance, and its weight can be measured. Depending on the form of matter of which they are made, some solid objects float in water while others sink. Many forms of matter are identifiable by their hardness or flexibility, by their taste and odor, or by the sound or light that they emit.

All things that we can sense directly are made up of progressively smaller things. Tools like microscopes, telescopes, or thermometers are used to observe things which are beyond the range of human senses. Standardized tools like rulers, scales, and clocks help people measure and communicate their observations with others. The more that is known about the properties of matter, the better people can describe its structure and construct an understanding of the world.

Activity 1: Oo-bleck—The Mystery Matter

This activity is derived from Activities to Integrate Mathematics and Science, 1987. Review with students the properties of matter (solids and liquids), then present them with the mystery compound (made with cornstarch and water).

Facts about solids

    do not change shape easily

    will not allow another solid to pass through them easily

    are usually visible

    have a definite shape

    when heated become a liquid

    when cooled stay solid

Facts about liquids

    change shapes easily (take shape of container)

    will allow a solid to pass through easily

    may be visible or invisible

    have a definite shape

    when heated become a gas

    when cooled become a solid

Invite students to test the mystery matter in various ways

    Can they pour it?

    Can they poke it; does it dent?

    Can they shatter it?

    Can they twist it?

    When heated over a candle does it melt?

Evaluation

    1. Have students graph the results of the tests. Using a circle graph, color in a section for each test result. Use one color for liquid test results, another color for solid. If a result falls into two categories, fill in half with each color.

2. Have students write a response to this question in their science and language-arts journal: According to the test results, is oo-bleck a liquid or solid? Why?

Themes: energy, systems and interactions, patterns of change.

Activity 2: Representing Atoms

The basic building blocks of all matter are very small particles called atoms. Atoms are much too small to be seen with ordinary microscopes, but they can be seen with recently invented atomic force microscopes. If a piece of matter was continually cut in half, a point would finally be reached where only a single atom would remain. Because atoms cannot be seen without special microscopes, this activity will let students look at the small parts of everyday things to get a feeling of what scientists mean when they talk about atoms.

Materials

one sugar cube, one four-color picture from a magazine, a small piece of fabric, one hand lens, pencil, science and math journals

Procedure

1. Examine a small portion of the picture. Now examine it again with a hand lens. Draw or write about what was observed in the science journal.
2. Examine the fabric, then use the hand lens. Again, record the observations in the science journal.
3. Examine the sugar cube, first without a hand lens, then with it. Draw or write a description of what was observed.

Evaluation

1. Compare the differences of the items seen with and without the hand lens.
2. Determine what the smaller parts of each of the items were made from.
3. Speculate what would happen if these items were examined under a standard microscope or under a microscope that could see atoms.

Themes: systems and interactions, patterns of change.

Activity 3: Cold Sodas Warming Up

Which will warm up faster?

A cold soda wrapped in tin foil or one wrapped in a wool hat?

The inside of the cap is at room temperature. It has not been just taken off somebody's head, as one clever student suggested.

How long will it take to bring the soda to room temperature?

1. Think about the problem.
2. Write down your reasoning.

3. Find a partner.
4. Discuss your reasoning based on what you know about math and science.
5. Come up with a graph.
6. Report your results to the class.

Homework assignment

Test it out. Do the experiment. Take the temperature every half hour.

The majority of some college classes and a few scientists, who know the theory, guess wrong on this one: So much for real-world application. The wool hat keeps the soda cold quite a bit longer. Though after about four hours the drink is at room temperature with both.

Background Information

One of the most important theories in physical science is the kinetic molecular theory: All matter is made of tiny particles, atoms and molecules. These particles are in constant motion. Though the idea is simply stated, it plays a fundamental role in chemistry, physics, biology, and geology. Our scientific understanding of heat and temperature, for example, is based on the kinetic molecular theory. It is impossible to understand respiration and photosynthesis, ecology or weather if you have no understanding of kinetic molecular theory. In other words, it is impossible to understand most of modern science without a good understanding of kinetic molecular theory. Translated to practical experience, this means that in most situations increased temperature means greater average energy of motion, so most substances expand when heated. In solids, the atoms are closely locked in position and can only vibrate. In liquids, atoms or molecules have higher energy of motion, are more loosely connected, and can slide past one another. Some molecules may get enough energy to escape into a gas. In gases, the atoms or molecules have still more energy of motion and are free of one another except during collisions (NRC, 1995).

Force, Friction, and Simple Machines

Force is the term given to pushes and pulls. Objects (both inanimate and animate) exert forces on other objects. The pull or attraction that each body has for another is called gravity. Gravity is the term given to the pull of the earth on objects. The earth's gravity pulls on everybody at or near the earth's surface. The downward pull of the earth's gravity holds the air and water on the earth and also keeps people from falling off the earth.

The measure of the earth's pull of gravity on a body is called the weight of that body. Weights (forces) are often measured with a scale. The customary units of force are pounds and ounces or grams and kilograms. The direction in which a force is exerted is important in determining its effect. If one object exerts a force on another object, the second object exerts an equal force in the opposite direction on the first. For example, in order to start to walk or run, a person must push with his or her foot against the floor in the direction opposite to that in which he or she intends to move. The floor pushes back against the foot. This unbalanced force is the direct cause of the change of the person's forward motion.

Friction is the force that resists the movement of one material over another material when the surfaces of two materials rub against each other. The nature both of the materials and of the surfaces affects friction. For example, firm hard materials produce less friction than materials that are soft or sticky. In like manner, less friction is produced by smooth exteriors than by rough surfaces. Friction can be reduced by making surfaces smoother or by placing a slippery material, called a lubricant, between two surfaces. Rollers, wheels, or ball bearings also reduce friction.

In many cases, friction is useful. As mentioned, friction helps us walk, and without friction we could not write, drive a car, open jars, throw a baseball, or light a match. But friction can also be harmful. Friction wears away materials like shoes, automobile tires, clothes, and machine parts. Friction also produces heat which may be harmful. For example, if a person falls or slides along the floor or carpet, the heat produced by the friction may cause a painful burn.

Energy, Heat, Sound, Light, Magnetism, and Electricity

Energy presents itself in many forms, including heat, light, and sound as well as gravitational and mechanical energy. Energy changes constantly. Despite this flux in energy gain and loss, we know that the total energy of all systems remains constant. This idea is known as conservation of energy. Energy may be stored or expressed in motion. Changes in motion, whether it is speeding up motion, slowing it down, or changing the direction, are due to the effects of forces. Two common forces are gravity and magnetism.

Energy can be classified in many ways, depending on the purpose of the experimenter. Energy is shown when we drop a ball, clap our hands, strike a match, turn on a light, or make waves in a bathtub. Each form of energy has its own characteristics. For example, a specific material will transfer some forms of energy and absorb or reflect others. A thick sheet of paper transmits sound but not light. Heat is often made by conversion from other forms, as demonstrated by the warming of a dark object by sunlight. Energy is required when work is done on a system or when matter changes its form.

The source of most of the energy people use is the sun. This renewable resource is supposedly unlimited (for practical purposes). Change in matter occurs because of the transfer of energy. In the physical sciences, energy can be explored in a variety of ways: heat, light, sound, electricity, magnetism, and so forth. Energy is an important theme to the physical sciences because all physical interactions involve energy.

Heat Energy

Heat energy is produced in many ways. It is created in an object when the object is exposed to the sun, to fire, or to other light sources. Heat energy can also be caused by rubbing two objects together. Heat moves from a hotter place to a cooler region.

The sun is the most important source of heat on earth. Plants need heat and light to make food. Animals, including people, need the sun to keep warm. Living things are most comfortable at certain temperatures. Temperature is the measure of the amount of heat in something.

The tool for measuring temperature is called a thermometer. An ancient word, thermometer is actually derived from two words: *thermo* means heat and *meter* means to measure. Thermometers are divided into units, called degrees.

Though heat and temperature are not the same thing, they are related to one another. Most materials expand with increasing temperature and contract when the temperature decreases. All substances are made of tiny particles, called molecules. These molecules are always moving. The movement of these molecules is called heat. Heat is a form of energy. It is the energy of moving molecules.

Light Energy and Color

Light arouses our sense of sight, considered by many to be the most important of our senses. Light is not something that we see or touch. In seeing, our eyes and brains are responding to the light that reflects from objects we are looking at. Light comes into our eyes, the part of the body that is sensitive to light. A picture of what we are looking at is formed in our brain. The eye can see in both dim and bright light, but there are limits. Looking directly at the sun is dangerous because the sun is such a bright source of light. Most objects do not make their own light but reflect light from other sources. Light is classified according to its brightness and color.

Light radiates in all directions outward from the sun and other hot materials. Light travels in straight lines, if not interrupted, and at tremendous speeds (186,000 miles per second). The speed of light appears to be the natural speed limit of the universe. Light is one way energy is sent from a source to the object that absorbs it. All external parts of objects reflect some of the light that hits them. We are able to see materials that reflect light as well as objects that make their own light.

Sound and Energy

Sounds are everywhere. Some come from nature, others come from machines. Many are made by people in the normal course of work or play.

Activity 1: Sound and Energy (compressing molecules)

Musical instruments make all kinds of sounds. In this activity, students are going to discover different ways to produce sound and try to determine how they are all alike.

Materials

rubber bands of various length and thickness

large paper clip

ruler

Procedures

1. Have students work in partnerships.
2. Instruct students to stretch one of the rubber bands and have their partner pluck it.
3. Encourage students to write what they observed in their science and math journals.
4. Have students repeat Step 2 using different rubber bands and record new observations.
5. Instruct students to bend the paper clip and hold it on the table.
6. Direct students to pluck the top end of the paper clip and observe what happens. Have them record their observations in their science and math journals.
7. Have students hold their ruler firmly against their desktop with about half of the ruler sticking out over the edge.
8. Have students pluck the free end of the ruler and record their observations.
9. Pose this challenge to the students: See if you can find a way to control the sound.

Teachers Note

The frequency of vibrations is higher or lower depending on the length or tightness of the rubber band. The faster the vibrations, the higher the frequency; the higher the frequency, the higher the pitch.

Evaluation

1. Have students write a description of what they made.
2. Encourage students to describe ways they controlled the sound of their instruments.
3. Instruct students to record other observations about each of the sound makers.

4. Have students reflect in their portfolio on what all of the objects had in common while they were producing sounds.
5. Ask students to think of other common objects that produce sound in the same way that the rubber band, paper clip, and ruler did. Ask students what they had to do to make the objects produce sound.
6. Have students include their reactions and what they learned from this activity.

Themes: energy, systems and interactions, patterns of change.

Discussion: What Causes Sound?

In the last activity you saw that when you plucked or moved some common objects they produced sounds. None of them could have produced sound on its own. You saw that the rubber band, the paper clip, and the ruler were moving back and forth rapidly in order to produce sounds. It is this same kind of rapid back-and-forth motion that causes all sounds. This kind of motion is called *vibration*. Clapping your hands, shouting, or closing your book are all examples of vibration. All these actions produce vibrations in the air; all these sounds take work. Energy must be transferred from one object to another. Can you now see how sound is a form of energy?

Activity 2: Testing the Speed of Sound

Materials

hammer

piece of wood

Procedure

1. Instruct students to go outside into a large open space.
2. Direct a student to place a thick piece of wood on the ground and instruct the rest of the class to walk a few steps away.
3. Direct another student to sharply hit the wood once with a hammer.
4. Ask the class located a short distance away if they heard the sound at about the same time the hammer hit.
5. Have the class move further away and direct the student to hit the wood again. Repeat several times, each time moving further away.

Evaluation

Have students record the distance and the time they observed the hammer hit the wood in their science and language-arts journals. It is helpful for students to count in their group by seconds.

Themes: systems and interactions, energy.

Magnets and Electricity

Today, magnets are all around us. That, however, was not always the case. Ancient mariners relied on crude compasses called lodestones.

Lodestones are made of magnetite, an iron ore found in various locations on the earth's crust. Only a few of these deposits are magnetized.

No one knows for sure who first used a magnetized needle on a pivot setting in a container to make a compass. Legend has it that the Chinese passed the idea on to the Arabs, who passed it on to the Europeans. Sailers were probably the first to use compasses. Around 1269, a French scholar, Peter Peregrinus, named the poles and described how they attract and repel. For a long time, scientists believed the earth must have a giant iron magnet at its core. Though the earth's core is iron, this cannot be a magnet. The key to the earth's magnetic properties appears to lie in the link between magnetism and electricity.

This link was first discovered in the 19th century. Hans Christian Oersted, a Danish physics professor, was giving a lecture on electricity. In showing how electric current heated wire, he was surprised to discover that an electric current has magnetic effects just like a lodestone or steel magnet. Many scientists followed this discovery, leading to induction (movement of magnetic lines of force to generate electricity) and opening the way for everything from electric lighting to telecommunications.

Static Electricity

Static electricity is produced by friction. Many people mistakenly think that lightning is an example of static electricity. Lightning, though caused by static electricity, is actually current electricity. When two different materials are rubbed together, they attract light objects like small bits of paper or pieces of thread. Why? An electric charge is an excess or deficiency of electrons in a substance: An excess of electrons produces a negative charge; a deficiency of electrons produces a positive charge. Because the object is static (not moving) the object itself becomes electric. When a charged object approaches a neutral object, the neutral object becomes charged by induction. At first glance it may appear that the law of electrostatic attraction and repulsion does not hold true as pieces of paper, thread, puffed wheat, or foam which have not been charged by friction are attracted to a charged balloon. Try this experiment for yourself.

Sample Activity: What Is Attracted to a Balloon?

Materials
    balloon
    wool, felt, or cotton
    small pieces of paper
    pieces of thread

small pieces of foam

puffed rice cereal

Procedure

1. Have students blow up a balloon and make a knot at the top.
2. Rub the balloon with the wool, felt, or cotton.
3. Instruct students to bring the balloon closer to the paper.
4. Have students write what happened in their science and math journals.
5. Next, direct students to predict what will happen to the other objects. In their journals have them set up a chart.
6. Encourage students to test their predictions.
7. Challenge students to find objects that are not attracted to the balloon.
8. Have students record what is and what is not attracted to the balloon charge.
9. Pose this question to the class: What could be done to make the balloon go back to a neutral state?

Evaluation

1. Instruct students to come up with rules of what is and what is not attracted to a static electrical charge.
2. Can students list some of the properties of the objects that were attracted to the charged balloon?
3. Have students write in their science and math journals what they learned from this experiment about static electricity.

Themes: energy, systems, and interactions.

Current Electricity

Scientists do not know exactly what electricity is, but they do know that electricity describes the flow of electrons. When an electric current is flowing through a material, the electrons move from atom to atom inside the material.

A simple electric circuit has three parts:

1. A source of electricity (dry cell or electric generator).
2. A path where the electric current can travel (copper wire).
3. Something that uses the electricity (light bulb, bell, machine, or other appliance).

In a simple circuit, the electric current flows from the source of electricity along one path to the appliance, passes through the appliance, and then returns through a second path to the source of electricity.

When all the parts of the circuit are connected so that an electric current is moving through it, the circuit is complete or "closed." When any of the three parts are disconnected and the current is not flowing, the circuit is incomplete or "open."

Activity 3: Bulletin Board Energy Test

This test is based on an idea presented by Robertta Barba in *Science in the Multicultural Classroom*, 1995.

Materials

    poster board or newsprint (one per group)

    magazines

    scissors

    felt-tipped markers

    crayons

    glue

Procedure

1. Have students design a bulletin board that shows their knowledge of energy.
2. Direct students to make sure that the display includes a definition of energy, and an impressive display of the types of energy.

Evaluation

The following rubric can be used to evaluate the groups energy posters:

Points	Characteristics
0	Shows no evidence of a knowledge of energy.
1	Group is able to define energy operationally.
2	Groups are able to define energy effectively and show a knowledge of at least three types of energy.
3	Students are able to clearly define energy and give examples with the skill or knowledge of an "expert."

MAPPING THE FUTURE

Teachers who enthusiastically adapt and use inquiry-based activities are likely to motivate their students to experience the beauty and power of scientific understanding. As teachers read, go to conferences, take classes, and work on improvement with colleagues, they are bound to develop a sophisticated set of approaches to teaching science. Of course, students must still be challenged to take responsibility for their own work. But as teachers model positive attitudes about science and its associates (math and technology), the result is bound to be infectious.

Interdisciplinary understandings are usually easier to grasp after students have a base of understanding in the subjects being connected. As teachers search for better ways to engage students, they look for naturally occurring links between the powerful ideas and organizing concepts that cut across disciplines. Many current programs and textbooks build on interdisciplinary themes, so the ideas, examples, and activities included here can be incorporated into a wide range of science or language-arts programs. These include efforts to connect student's learning with their experiences. The various elements fall at different places on a continuum of sophistication and complexity, depending on the teachers goals for science instruction.

Students now have to go beyond subject matter to understand the underlying interdisciplinary themes that hold the content together. Here we have outlined the suggested themes for science and language-arts learning. Earth, life, and physical sciences are connected through theme studies and activities to help students experience connections between these three domains. This is done in a manner that tries to retain the specific integrity of each content strand and suggests that specific elements of content can be taught more effectively by providing a context for the knowledge and skills students learn.

The potential benefits that stem from moving toward a quality science and language-arts education program are so great that we must act now. Changing how we go about teaching means taking a route that is difficult, time consuming, and deeply rewarding. Being literate in today's society includes being active, critical, and creative users, of print and spoken language, and also encompasses an individual's ability to decode, interpret, and analyze the unifying ideas of particular subject areas.

As teachers find natural (rather than forced) connections, content integration can help students apply their understanding of the natural environment, the world of language, and the nature of technology. Once an interdisciplinary theme has been identified, teachers can identify how the possible parts of the lesson relate to district, state, and subject-matter standards. This way, curriculum standards support the enterprise and give the curriculum more power in the eyes of parents, administrators, and other teachers.

It is important to remind ourselves that our visions do matter. As teachers become more knowledgeable and enthusiastic about integrated inquiry with science and the language arts, they will be able to act on their highest visions and map the future of science, technology, and language-arts education.

The visions we offer our children shape the future.
—Carl Sagan (1994)

Chapter 8

Educational Technology: The Multiple Possibilities of Powerful Tools

We are, all of us, being drawn into the electronic world, and we can't stop it. It's like being given a car without any one telling you how to drive it, and you don't have a road map. We're driving blind.
—Sandy Sparks (1997)

To some degree, communication and information technology has always reflected the strengths and weaknesses of the human condition. The technological tools of today can serve as passports to an ever-expanding reservoir of knowledge and rapid human communication. But they have to be tailored for a context that is more or less appropriate.

We are just awakening to the fact that the computer and the World Wide Web are redefining literacy and reshaping how we learn. Just as Gutenburg made possible the stories that ushered in the modern era, the digital medium that is growing up now will determine the nature of more interactive stories in the 21st century. Clearly, technology is in the process of redefining how we learn, play, and understand our lives. And just as clearly, the ability to understand, evaluate, and integrate information delivered by electronic media is becoming an important part of science, language, and literacy.

The technologically intensive society of tomorrow will bring with it profound shifts in everything from the mass media to the classroom. Powerful sensory presence and participatory formats come with the territory. There are intellectual and moral consequences to being able to tailor the information that reaches us. Image and instant reaction can, for example, overwhelm meaning and careful deliberation. Without much thought or assessment we embrace new technologies that amplify the production of data. We are now awash with e-mail, TV news, advertising, infotainment, and trash on the Internet. Instead of information scarcity, we have created a new problem: information glut and incoherence. A colossal amount of time can be wasted. It is time to create intelligent possibilities and hopeful visions.

As technology has become more powerful and intrusive, our society has become infatuated with it. For example, the public and the politicians seem sure that simply connecting the schools to the technological juggernaut will take care of fundamental educational problems. Good luck. As many teachers will tell you, technology alone is not enough, there has to be a pedagogical plan. Educational technology only works well when it is driven by instructional strategies and a quality curriculum. Once clear educational goals and subject-matter standards are in place, technology can enable important changes in the curriculum and even help the schools move down the reform trail.

CONSIDER THE MYTHS AND THE MAGIC

Where technology is taking us remains something of a mystery. Some of the consequences of technology can be predicted; many cannot. For example, who in the early 1970s could have predicted where computer technology was going to take us in the 1980s? The liberation of human potential or destruction of privacy? Both, actually. Technology shapes and reflects the values found in society. At school, it can isolate learners or help them join with others. In our personal and civic lives, technological tools frequently slip through our hands to limit our choices at work and erode the edges of the Constitution in our daily lives. Heaven help those who do not get caught up in it. Do you dare to question the time-wasting computer-based functions that dominate your life? If you cannot program your VCR or answer hundreds of less-than-private e-mail messages, just wait until you see what techno-enthusiasts are going to bring you next. We suggest injecting a little healthy skepticism into the debate. Consider the myths along with the magic.

In spite of some misplaced enthusiasm, computers and connections to the Internet are technological assistants that are now being used in many schools. By the mid-1990s, it was clear that the Internet was not just a temporary fad of the pocket-protector set. Now lessons can take on a whole new life if they include the Internet. Fast-paced and stimulating, it can be used for investigation, analysis, inquiry, problem solving, and communicating with people halfway around the globe. Let us hope that the artificial stimulation offered by new electronic worlds will ultimately cause people to value human company more. Helping students see the human factor and the societal implications is important if technology is going to do something worthwhile for our culture and our society.

Though some benefits are undeniable, it will take much more than technology to clearly improve instruction or fundamentally reform American education. Whether it is schooling or technology, progress is usually incremental. Spectacular new approaches and theories are rare events and will continue to be so. The key is professional development, the content of technology-assisted lessons, and how both are connected to what is going on in the classroom. National standards that include technology have helped, and it is important to keep new subject-matter standards in mind as we move on integrating technology-based possibilities into daily lessons.

THE STANDARDS AND TECHNOLOGY

The science standards (NRC, 1995) view technology as helping to form connections between the natural and man-made worlds. The laws of the physical and biological universe are viewed as important to understanding how technological objects and systems work. The standards for the English language arts (NCTE, 1996) also refer to visual literacy, information, and communications technology as central factors in learning language and literacy. Both sets of standards suggest that students should be given opportunities to use technology to access information from around the world as they explore and design solutions to problems.

The problem-solving ability of children is helped by using technological tools the same way adults use them. Student-generated problems provide excellent opportunities to direct attention to some of the new tools and instruments used in science and the language arts. In early elementary grades, many tasks can be designed around the familiar contexts of the home, school, and community. In the primary grades, it is usually best when there are only one or two possible solutions that do not require a great deal of preparation time or complicated assembly.

At all levels, multidisciplinary analysis of problems is natural when it connects to the students' day-to-day world. A sequence of five stages are usually involved in a technology-based problem-solving process: (1) identifying and stating the problem, (2) designing an approach to solving the problem, (3) implementing and arriving at a solution, (4) evaluating results, and (5) communicating the problem, design, and solutions. Students can also design problems and technological investigations which incorporate several interesting issues in science, language, and literacy. By using a variety of technologies for real-world inquiry, problem solving, and communicating, students can come to recognize that learning is more than preparing for life; it is life itself. As in life, when on-line it is best to respect others, be generous, and not get into silly fights.

CONNECTING TECHNOLOGY TO CONCRETE REALITY

The Egg Drop

The first example is a design activity where students design and test a container that can keep a raw egg from breaking when dropped from the ceiling (about 8 feet). There are computer programs that can aid in the design and the Internet has a wealth of information. The activity may be preceded by a science unit on force and motion so that students are able to apply their knowledge of science in their design process. Students can use an Internet search engine to find out about the possibilities. They can also communicate with students who have tried the experiment. Students will work in small groups in planning their egg drop design. Emphasize creativity. Students are to bring materials from home to finish their design. If that problem is too difficult, have students design a container that is an egg catcher.

Procedure
1. work with a group
2. brainstorm ideas
3. sketch a design
4. formulate a rationale
5. assign group tasks, including clean-up crews
6. get materials (string, paper towel rolls, styrofoam peanuts, cotton, soft packing material, etc.)
7. build the container
8. try several tests
9. do a class demonstration

The presentation will be started with a discussion of what the group did to meet the challenge. Assessment for the egg drop is not whether the egg broke, but rather how they were able to share what they found out as they tried to solve the problem and prepared for a successful attempt. Making a video of the demonstration may add an interesting element. It can be viewed again by the designers or by parents, or used in other class sessions in years to come.

Designing a Model City

Another interesting problem for middle school students is to design and build a city. There are many simulation programs that can help with this. Students are instructed to design a city with an efficient road network. They must also create an election process which ensures that the city council fairly represents all city residents. In addition, students must contact construction companies and make a plan for building their cities. To prepare for this challenge, students should learn about routing graphs, which are used to plan routes for mail carriers and garbage carriers so they do not waste steps or gas unnecessarily. Contractors also use routing graphs to plan roads in new residence communities. Students collaborate in groups, analyzing their decisions by writing a rationale for their design decisions. They must also make a fifteen-minute oral presentation to "sell" their cities. This project allows students to be creative in applying the science, math, and technology applications they have learned. Some students have created their cities on islands, on the moon, even underground.

Communications Time Line

Time is often a difficult concept for children to grasp. Throughout modern history, people have recorded the passage of time. This activity gets students involved in time measurement by using a number of old and new technological tools and shows how the ways in which people communicate with each other have changed throughout history.

Procedure
1. Have students research the history of communications technology and create a time line in their science and language-arts journals.
2. Using actual objects or their representations, encourage students to assemble a communications time line project for display.
3. Remind students that each time period needs to have some examples of the actual objects used and a written explanation about these communications devices.

Comparative Analysis

Ask students to go to the library and collect copies of ads from each decade of this century. Much of this may be on microfilm. Compare and contrast the images and relate them to social events in each decade. Ask students to produce their own ads using the production techniques they see pictured at a particular time. Use the style of the period.

Research can also be done on the history of cigarette advertising, misleading claims, and ad campaigns directed towards a specific audience. Zero in on who is responsible for an ad. Sources such as *The Standard Directory of Advertisers* (National Registry of Publishing, 1998) will tell you the name and address of the advertising agency who created the ad. Ask students to choose an advertisement that they think is offensive, irresponsible, or one that they think represents things fairly. They can write a letter of complaint or congratulations on a job well done.

Connecting Students to One Another

The role of educational technology is changing how science, language, and literacy is taught by changing the instructional environment and providing opportunities for students to create new knowledge for themselves. Computer-based technology can serve as a vehicle for inquiry-based classrooms. The Internet can give students access to data, experiences with simulations, and the possibility for creating models of fundamental science, language, and literacy processes. Educational technology has many elements: Computers, multimedia, the Internet, camcorders, television, and virtual reality make up just a partial list. New technology is transforming the social and educational environment before we have a chance to think carefully about why we want to use the technology and what we hope to accomplish with it.

The reports that over half the computing power ever produced has been produced over the last three years is true. It is also true that the dollar sales of personal computers have surpassed those of color televisions, the world's most ubiquitous consumer-electronics product. No matter how you do the count, the digital technology at the heart of computers is moving into television sets, VCRs, and systems that link the telephone to the television. Digital devices manipulate data as electronic pulses represented by the 1s and 0s of computer code. Increasingly, they are converting text, pictures, sound, and video into a form that can be recorded, played back, transmitted, and received. Like everyone else, teachers are consumers of technology and they need to be able to judge critically the quality and usefulness of the electronic possibilities springing up around them. This requires continuing high-tech inservice training and attention to the education of new teachers.

Students and teachers need minimal digital competencies in order to use classroom computers as tools. In one technologically savvy sixth-grade classroom we visited, students were involved with software evaluation, spreadsheet applications with *Claris Works*, e-mail merge letters, interactive writing with *Prompt Writer*, making posters with Broderbund's *Printshop*, multimedia-based reports with *Hyperstudio*, and navigating the Internet with *Netscape*. The homework question of the week was, "What is the role of media in our society?" Not many of us could juggle all this and integrate the results into the curriculum.

This chapter should help teachers less familiar with technology consider some of the general issues and make the appropriate match between the problems they face and potential technological support. As the standards in science and math make clear, children can learn a great deal about both subjects from the high-tech and the low-tech end of the technology spectrum.

CHANGING TECHNOLOGY BEFORE IT CHANGES US

Computers are the most powerful recent example of a technology being promoted in our schools before many teachers have a clear understanding of what might be accomplished and what the benefits might be. Since the early 1980s, schools have been caught between aggressive marketing, technological myths, and promising possibilities. The professional development of teachers and the quality of the software were often neglected, even though they were key ingredients for integrating technological tools into their classrooms. Teachers who know how to structure active inquiry-based learning have little difficulty using the computer as a powertool for such things as visually exploring models and problem solving.

What are some technological negatives that we should keep in mind? To begin with, people often uncritically accept the parameters of computer programs, even when the simulated environment is very wrong. Multimedia compounds the problem of uncritical consumers. We are on the verge of losing more than we know. It is often easier to attend to a television or a computer than it is to make meaning from print. But simply being motivated by a computer program or the Internet does not mean that students are learning something important.

In our new media-fed society, images can engage public attention with small controversies and trivial banalities. However, this same media-connected world can also provide students with the possibility for controlling and charting the course of their education and their culture. Information can now be constructed by anyone with a computer, camcorder, and Internet link. And the best of the new technology moves you out of a passive realm and into interaction with others.

New ways of relating to electronic information requires a break from habit. Thousands of years ago it was the written word. Next it was the printing press. Today it is multimedia computing, the coming together of computers, video, sound, animation, and telecommunications. Computers are both evolutionary and revolutionary. At their best they help you do things (like typing) better, while conjuring up new possibilities for critical thinking, collaboration, and creativity. As we struggle with school reform we need all the help we can get. Multimedia computing adds a new dimension to learning by communicating meaning with vivid-motion video, animation, and quality sound. Some multimedia workstations even have built-in video cameras so that users can hold video teleconferences with people in distant locations and construct their own video compositions. By motivating students through the excitement of discovery, technology can assist the imaginative spirit of inquiry and make lessons sparkle.

Distance learning is taking off for a bumpy ride in our new electronic environment. There is a major expansion of the kinds of services and courses that are being provided through distance learning, satellites, and networks like the Internet. Some of these efforts supplement classroom activities and others are aimed at those who are seeking alternatives to classroom attendance. New systems are much less passive and allow distant teachers to illustrate course materials with sound, high-quality graphics, animation, full-motion video, and interactive problem solving and simulation.

USING MULTIMEDIA TECHNOLOGY TO OPEN DOORS FOR LEARNING

To be valuable, educational technology must contribute to the improvement of education. Technological tools should be designed to help open doors to reality and provide a setting for reflection—making important points that might otherwise go unnoticed. For example, computers can use mathematical rules to simulate and synthesize lifelike behavior of cells growing and dividing. The CD-ROM storage medium used for such programs has expanded to the point where there are now many thousands of titles for home and school use. Multimedia computers build in large amounts of text, excellent graphics, video clips, and stereo sound. It is possible to set the computer to the section of the video you wish to view, play that portion, and bring up print describing the imagery. If you want to play it again, or switch to another part of the disk, you can do it with a few strokes on the computer keyboard.

A recent improvement of the CD-ROM is Sony's DVD optical disk, which looks very much like a CD-ROM. The DVD optical disk can store up to 133 minutes of video material or, in terms of data storage, hold up to 4.7 gigabytes (billion bytes) of data—seven times the capacity of a typical CD-ROM. Sony's DVD is also virtually indestructible—it can be replayed hundreds of times without picture degradation, allowing random access to any segment of the video show while maintaining superbly sharp picture and color clarity.

Progress in school reform will be limited if teachers are forced to use traditional instructional delivery tools or models. "Ask and tell" and "tell and ask" are limited patterns, with or without the computer. The interactive design of new multimedia programs are profiting from findings regarding cognitive development and collaborative learning. Computer-based activities can have problem-centered structures that wrap learning experiences around problems in new ways. You can visually enter the body as a blood cell or explore another planet. As budding scientists, students can use computer tools to visually explore empirical claims and examine all kinds of evidence that allows them to support or critique important findings.

As new technologies and related products start to fulfill their promise, students will become active participants in knowledge construction across a variety of disciplines—providing a technological gateway to learning in the 21st century. As state-of-the-art pedagogy is connected with state-of-the-art technological tools, it will change the way knowledge is constructed, stored, and learned.

Studies confirm that the power and permanency of what we learn is greater when visually based mental models are used in conjunction with the printed word. Inferences drawn from visual models can lead to more profound thinking. Children learn to rely on their perceptual (visual) learning even if their conceptual knowledge contradicts it. Even when it runs contrary to verbal explanations or personal experience, the video screen can provide potent visual experiences that push viewers to accept what is presented.

EXTRACTING MEANING FROM THE VIDEO SCREEN

One of the most accessible sources of electronic information is found in almost every home and classroom. Quality television can support student learning and interests. Unfortunately, it rarely does. The advocacy group Children Now released an analysis of 1995 programming that found that in the virtual (unreal) world of television, children rarely have real-world problems. They usually live carefree lives of affluence, have obscure family ties, rarely have pleasant learning experiences, and usually do no homework. If school is shown at all, it is as a backdrop. In addition, physical aggression usually pays off better than showing affection or meeting responsibility.

One way to help students become more intelligent video consumers is to learn to "read" and "write" with video. Children can become adept at extracting meaning from the conventions of video production—zooms, pans, tilts, fade outs, and flashbacks. With cheap camcorders, they can actually do some of this themselves. Distinguishing fact from fiction can be difficult with passive television. What we perceive falls within the framework of concept formation. Like print, visual imagery from a TV screen can be mentally processed at different levels of complexity.

Piaget (1973) showed how certain notions of time, space, or morality are beyond children's grasp before certain developmental levels are reached. Research on TV viewing suggests that it is not vocabulary limitations alone which impede children from grasping some adult content (Dede, 1985). Children lack fundamental integrative capacities to "chunk" (group) certain kinds of information into meaningful groups which are obvious to adults. Thus, children who need help in developing strategies for tuning out irrelevancies may be especially vulnerable to unwanted adult content (McKibben, 1992).

Most of the time children construct meaning for television content without even thinking about it. They attend to stimuli and extract meaning from subtle messages. The underlying message of most TV programming is that viewers should consume as much as possible while changing as little as possible. How well television content is understood varies according to similarities between viewers and content, viewers' needs and interests, and the age of the viewer. Sorting through the themes of mental conservatism and material addition requires carefully developed thinking skills. Meaning is constructed by each participant at several levels.

Broadcast television has provided us with a common culture that feeds on cynicism and selfishness. TV responds to the public hunger for community with programming that applies capitalist market values and standards to all human relationships. Literature is just one way to temper this with a little compassion and spirituality, and a few transcendent ideas. Building a culture of meaning is an increasingly difficult and varied undertaking. When students share an informed set of analytical tools for interpreting signals from media, the prospects for our public life will improve.

The greater the experiential background, the greater the understanding. The ability to make subtle judgments about video imagery is a developmental outcome that proceeds from stage to stage with an accumulation of viewing experience. Thus, different age groups reveal varying levels of comprehension when they view TV programs. Eight-year-olds, for example, retain a relatively small proportion of central actions, events, or settings of typical programs. Even when they retain explicit content, younger children fail to infer interscene connections. Improvement in comprehension occurs with maturity. But substantial understanding of the medium requires training for parents, teachers, and children. Training for parents may involve how to interact with their children, how to critique, analyze, and discuss what is viewed, and how to model good viewing habits.

Lessons structured around the TV medium can assist students in becoming intelligent video consumers and help them evaluate the multimedia medium that relies so heavily on video clips. Many public television stations are now offering special workshops for teachers in how to use educational television programs effectively in their classrooms. Some ideas include selecting short segments from programs, designing carefully crafted questions, after viewing getting students to talk together about what they watched, turning the sound off, replaying the tape again, encouraging students to come up with two or three things they found out from the short clip and sharing that information with others. Have students look for and identify propaganda techniques, act them out, and make up some skits (ads) of their own. Evaluation of television viewing is also an important comprehension skill. Encourage students to analyze and rate what they view.

Reflective thought and imaginative active play are important parts of the growth process of a child. Contrary to popular belief, children must do some active work to watch TV, make sense of its contents, and utilize its message. Evaluative activities include judging and assigning worth, assessing what is admired, and deciding what positive and negative impressions should be assigned to the content. In this sense, children are active participants in determining television's meaning.

Traditional television is vulnerable to more invigorating interactive electronic competition. Children will often pass up passive television viewing, for example, when they can use a computer or game controller to interact with the medium. Many children, for example, are already familiar with "Where in the World Is Carmen Sandiago." They learn geography by electronically rushing around the world and trying to catch up with Carmen. When the computer-controlled version goes up against the linear PBS television program of the same name, the interactive version usually wins hand down because of higher levels of personal involvement.

People learn best if they take an active role in their own learning. Relying upon a host of cognitive inputs, individuals select and interpret the raw data of experience to produce a personal understanding of reality. Ultimately, it is up to each person to determine what he or she pays attention to and what he or she ignores. How elements are organized—and how meaning is attached to any concept—is an individual act that can be influenced by a number of external agents. The thinking that must be done to make sense of perceptions ultimately transforms the real world into different things for different people.

COMPREHENDING MEDIA MESSAGES

Parents, teachers, and other adults can significantly affect what information children gather from television. And the skills learned from analyzing this visually intensive medium will apply to more advanced multimedia platforms. Students' social, cultural, educational, and family contexts influence what messages they take from the television, how they use TV, and how "literate" they are as viewers. To become critical viewers who are literate about media messages students should be able to do the following:

- Understand the grammar and syntax of television as expressed in different program forms.
- Analyze the pervasive appeals of television advertising.
- Compare similar presentations or those with similar purposes in different media.
- Identify values in language, characterization, conflict resolution, and sound or visual images.
- Identify elements in dramatic presentations associated with the concepts of plot, storyline, theme, characterizations, motivation, program formats, and production values.
- Utilize strategies for the management of duration of viewing and program choices.

Parents and teachers can engage in activities that affect children's interest in televised messages and help them learn how to process video information. Good modeling behavior, explaining content, and showing how the program content relates to student interests are just a few examples of how adults can provide positive viewing motivation. Adults can also exhibit an informed response, point out misleading TV messages, and take care not to build curiosity for undesirable programs.

The viewing habits of families play a large role in determining how children approach the medium. The length of time parents spend watching television, the kinds of programs viewed, and the reactions of parents and siblings toward programming messages all have a large influence on the child (McLeod, 1982). If adults read and there are books, magazines, and newspapers around the house, children will pay more attention to print. Influencing what children view on television may be done with rules about what may or may not be watched, interactions with children during viewing, and the modeling of certain content choices.

Whether coviewing or not, the viewing choices of adults in a child's life (parents, teachers, etc.) set an example for children. If parents are heavy watchers of public television or news programming, then children are more likely to respond favorably to this content. Influencing the settings in which children watch TV is also a factor. Turning the TV set off during meals, for example, sets a family priority. Families can seek a more open and equal approach to choosing television shows—interacting before, during, and after the program. Parents can also organize formal or informal activities outside the house that provide alternatives to TV viewing.

It is increasingly clear that the education of children is a shared responsibility. Parents need connections with what is going on in the schools. But it is teachers who will be the ones called upon to make the educational connections entwining varieties of print and visual media with science, mathematics, or technology. It is possible to use the TV medium in a way that encourages students to become intelligent video consumers.

BECOMING AN INTELLIGENT CONSUMER OF MASS MEDIA

To understand media messages, teachers can capitalize on the information and knowledge that students bring to class. It is important to look more closely at the business side of the media. From all the selling and salaries involved, it is clear that the mass media is an economically valued storyteller in our society. Some questions to discuss on this issue are the following:

1. What is your favorite website, TV show, and movie?
2. What kind of information, show, or movie are they?
3. What are the formal features of your choices?
4. What are the most appealing elements of each?
5. What do you know about how each media constructs their stories?
6. What are some of the formal and informal structures of the Internet, the movie industry, and TV broadcasting?
7. What are the values in these mass produced "programs" and how do they change our shared experiences as a people? (Fifteen years ago our shared experiences included books and newspapers. Now only the more educated members of our society do a great deal of reading.)

Help Students Critically View What They Watch

Decoding visual stimuli and learning from visual images requires practice. Seeing an image does not automatically ensure learning from it. Students must be guided in decoding and looking critically at what they view. One technique is to have students "read" the image on various levels. Students identify individual elements and classify them into various categories, then relate the whole to their own experiences, drawing inferences and creating new conceptualizations from what they have learned. Encourage students to look at the plot and storyline. Identify the message of the program. What symbols (camera techniques, motion sequences, setting, lighting, etc.) does the program use to make its message? What does the director do to arouse audience emotion and participation in the story? What metaphors and symbols are used?

Compare Print and Video Messages

Have students follow a current event on the evening news (taped segment on a VCR) and compare it to the same event written in a major newspaper. A question for discussion may be how the major newspapers influence what appears on a national network's news program. Encourage comparisons between both media. What are the strengths and weaknesses of each? What are the reasons behind the different presentations of a similar event?

Evaluate TV Viewing Habits

After compiling a list of their favorite TV programs, assign students to analyze the reasons for their popularity. Examine the messages these programs send to their audience; do the same for favorite books, magazines, newspapers, films, songs, and computer programs. Look for similarities and differences between the media.

Use Video for Instruction

Using a VCR, make frequent use of three- to five-minute video segments to illustrate different points. This is often better than showing long videotapes or a film on a videocassette. For example, teachers can show a five-minute segment from a videocassette movie to illustrate how one scene uses foreshadowing or music to set up the next scene.

Create a Scrapbook of Media Clippings

Have students keep a scrapbook of newspaper and magazine clippings on television and its associates. Paraphrase, draw a picture, or map out a personal interpretation of the articles. Share these with other students.

Create New Images from the Old

Have students take rather mundane photographs and multiply the images, or combine them with others, in a way that makes them interesting. Through the act of observing, it is possible to build a common body of experiences, humor, feeling, and originality. And through collaborative efforts students can expand on ideas and make the group process come alive.

Use Debate for Critical Thought

Debating is a communications model that can serve as a lively facilitator for concept building. Taking a current and relevant topic and formally debating it can serve as an important speech and language extension. For example, the class can discuss how mass media can support political tyranny, public conformity, or the technological enslavement of society. The discussion can serve as a blend of social studies, science, and humanities study. You can also build the process of writing or videotaping from the brainstorming stage to the final production.

Include Newspapers, Magazines, Literature, and Electronic Media in Daily Class Activities

Use of the media and literature can enliven classroom discussion of current conflicts and dilemmas. Neither squeamish nor politically correct, these sources of information provide readers with something to think and talk about. And they can present the key conflicts and dilemmas of our time in a way that allows students to enter the discussion. These stimulating sources of information can help the teacher structure lessons that go beyond facts to stimulate reading, critical thinking, and thoughtful discussion. By not concealing adult disagreements, everyone can take responsibility for promoting understanding. This engages others in moral reflection and provides a coherence and focus that helps turn controversies into advantageous educational experiences.

CHOOSING COMPUTER SOFTWARE

Most teachers subscribe to a number of professional journals and just about every school staff room has dozens. Some of the journals can be given to upper-grade students so that they can help with the selection. Check for reliable reviews on the Internet. Many contain software reviews that can keep you up-to-date. Magazines like *Electronic Learning* publish an annotated list of what their critics take to be the best new programs of the year. Even some of the old reliables from the 1980s (such as National Geographic and Nova programs) have been improved with sound and video and put on CDs. Also, district supervisors often have a list of what they think will work at your grade level. Of course, you can get your class directly involved in the software evaluation process. This helps your students reach the goal of understanding the educational purpose of the activity. We like to start our workshops by having teachers work in pairs to review a few good programs that most school districts actually have. Three examples are *Number Connections* (Sunburst), and, for the upper grades, *Fossil Hunter* (MECC) and *Operation Frog* (Scholastic). As you and your students go about choosing programs for the classroom the following checklist may prove useful:

1. Can the software be used easily by two students working together? (Graphic and spoken instructions help.)

2. What is the program trying to teach and how does it fit into the curriculum?
3. Does the software encourage students to experiment and think creatively about what they are doing?
4. Is the program lively and interesting?
5. Does the program allow students to collaborate, explore, and laugh?
6. Is the software technically sophisticated enough to build on multisensory ways of learning?
7. Is there any way to assess student performance?
8. What activities, materials, or manipulatives would extend the skills taught by this program?

The bottom line is if you and your students like it. We suggest that teachers reserve final judgment until they observe students using the program. Do not expect perfection. But if it does not build on the unique capacities of the computer, then you may just have an expensive electronic workbook that will not be of much use to anybody. With today's interactive multimedia programs, there is every reason to expect programs that can invite students to interact with other people, information, and phenomena from the biological and physical universe. Students can move from the past to the future and actively inquire about everything from experiments with dangerous substances to simulated interaction with long-dead scientists. Just do not leave out experiments with real chemicals and experiences with live human beings. Remember, actually touching, manipulating, and working on authentic things beats any computer-based trick.

INTERNET TECHNOLOGY AS A PARTNER IN LEARNING

New technology has always posed new dilemmas for parents and teachers. The printing press took nearly fifty years to perfect. High technology in moving in at warp speed. How much does it increase productivity? No one really knows, as the research on effectiveness is cloudy and sparse at this point. For example, few had even heard of the Internet five years ago. We only have some anecdotal evidence on its effectiveness in schools.

In many respects, the Internet is just another communications medium. It is a worldwide network of computers, usually connected by telephone lines. Some computers connected to the Internet are called "servers." They store information that can be retrieved from a home or school computer (called a "client"). When you download information from the Internet, your computer is getting information from a server. The World Wide Web consists of connected servers and clients that communicate using a common set of tools (a Web browser) like Netscape's *Communicator* or Microsoft's *Internet Explorer*.

The competition between Netscape and Microsoft means that a single browser is no longer a window onto the entire Web. Both companies are rapidly adding possibilities for video, audio, and more interactive information. To fully use all of the Web means loading both Netscape's *Navigator* and Microsoft's *Explorer* onto your PC until things coalesce around a single standard.

Technology magnifies human strengths and weaknesses. Television brought a new visual world into the living room. It was tame, but it did not do much for enlightenment. The Internet is different because it is not passive nor filtered for appropriateness. The interactive presence of unknown outsiders opens up the possibility of fraud, child abuse, and harassment, to say nothing of other aggravations. It also gives users the ability to be heard across the world and provides the possibility of finding information about almost anything. But how do you sort out what is valid from what is foolish? Have students find something on the Internet that is funny and clearly false, something that is without question true, and something that could be taken either way. On the foolish side, for example, my students recently found a website that maintained that one in three of us is a werewolf.

Coupling free expression with social responsibility is difficult at best. As libraries place filters (like *Netscape Nanny*) on all of their multimedia computers, adults have their choices circumscribed. Just try getting visual information on an important painting like "Nude Descending a Staircase," or even Super Bowl XXII. Anything with an "X" or the word "nude" is out of the question with some of these "child protection" programs.

The following are some suggestions for keeping children away from sordid content:

- Put computers in a location where an adult can see them.
- Remind students that
    People sometimes misrepresent themselves on the Internet.
    They should never respond to belligerent or suggestive messages.
    They should not give out home addresses, personal pictures, or telephone numbers.
    They should not download from unfamiliar sources.
    Acquaintanceships made on the Internet should stay there.
    They should work in pairs with at least some adult supervision.

What about sites that promote racial discrimination? When hate spills onto the Web the debate over whether or how to control it heats up. We have yet to figure out how to balance two of the most powerful revolutions of the 20th century, those of human rights and information technology. Regulations bring practical, legal, and technological problems with them. It may simply not be possible to ban hate sites and pornography without infringing on free speech.

Colleges and universities have long prized the right to explore any subject in the pursuit of knowledge. Freedom of expression is accorded the highest respect. Public or private, institutions of higher education try to give groups, even those on the margin, a forum for discussion. Still, we as a society must continue to make distinctions between what is and what is not appropriate for children. How can we do this without violating the First Amendment or reducing everything to the intellectual level of a seven-year-old?

Students can communicate with students in other countries on the Internet. The anecdotal evidence suggests that it can encourage reading and writing. On the other hand, when I walk into the college computer lab and watch students chat, surf, and make dates it is hard to come away without feeling that they might as well be watching daytime TV. How can we go about creating an instructional atmosphere that is enabled by the technology?

Computers and their associates are rarely used to compute anymore. They are used to communicate. Everyday students at all levels are receiving and sending large volumes of text, images, and sound. Just about anyone can construct a Web page. These electronic messages have a much wider audience than any of the more traditional forms of communication. One of the major goals for any educational innovation is to make subjects more interesting, comprehensible, and connected to everyday life. This means connecting to what the curriculum is trying to accomplish and looking for lifelong learning possibilities.

High-tech companies see a combined opportunity to create a market and possibly do a little social good. Much of this is driven by a dream of magical solutions to our educational problems. In spite of this cyber-romance, the positive effects of technology are questionable. We are not really sure what works and what does not. In spite of calling for a little skeptical inquiry, those of us who have used it believe that technology can support real-world concepts with visual imagery—helping students actually see problems. In addition, something very special happens when students know that their research (images and print) is available to other students around the world. The same can be said of the special appeal of bringing images, text, and sound from the outside world into isolated classrooms.

It is important to remember that learning is more about mental models than it is about imagery. Hooking schools up to the Internet has captured the imagination of many. Over 70 percent of the nation's schools now have at least one computer connected to the Internet. Does the Internet inform learning? Most connections have only been in place for a year or two. Like many aspects of educational technology, the jury is still out on the educational value of being on-line. Techno-enthusiasts maintain that great benefits are about to be realized. Critics say that the whole thing is a colossal waste of time. The truth is someplace in-between.

Multimedia computers go well beyond text to video, audio, and photos. This information is multidimensional and interactive. Being able to evaluate the source is more important than ever.

Skeptical Inquiry and the Internet

1. Give students a one-page scavenger hunt activity where they have to find everything from the atomic number of uranium to DNA research to how to say "hello" to people in Finland.
2. Find some current information that you believe. Explain. Find something about current events that you do not believe. Explain.
3. Examine the validity of something that you find.
4. Critique a website.
5. Go through some websites and ask some questions about the content. Come up with some answers. Learn to ask better questions.
6. How would the accuracy of information from an advertisement be compared to information found on a site like that of National Geographic?

A "Found" Lesson That Starts and Ends with the Internet

Like many good ideas on the Internet, we do not know who to credit for a lesson that many of us kept adding too. We changed it for gifted students in a middle school. First, we explored the Internet for pictures and print of early microscopes and set about actually making them much the way Dutch scientist Anton van Leeuwenhoek did in the 17th century. We melted glass into tiny beads to serve as lenses and put them in place between thin pieces of metal. Once this was done, we used the primitive device to view algae and other specimens. This hands-on activity went back to the Internet for music and art of the time. Next, the students found information about the role microscopes played in science, literacy, and philosophy. The Internet was used to help students understand what it was like to live in those times. Colored pictures were printed out to illustrate a journal. The whole experience was placed in a "sketch journal" that explained the whole experience in the form of a three- or four-page comic book.

The following are a few of the websites we use:

Resources at the Smithsonian
http://www.si.edu/resource/start.html

The San Francisco Museums of Fine Arts
http://www.thinker.org

Treasures of the Czars
http://www.times.st-pete.fl.us/Treasures/Default.html

The Louvre
http://www.Louvre.fr

Metropolitan Museum of Art
http://www.metmuseum.org

The Discovery Channel School
http://school.discovery.com

The Art Institute of Chicago
http://www.artic.edu/aic/firstpage.html

Some Hot Bookmarks
http://www.rice.edu/armadillo/Rice/Hotlinks/museums.html

Exploratorium Home Page
http://isaac.exploratorium.edu

AskERIC (askeric@askeric.org)
http://ericir.syr.edu

ERIC Clearinghouse on Reading, English, and Communication
http://www.indiana.edu/~eric_rec

Yahoo Index to the Regions of the World
http://www.yahoo.com/RegionalRegions

KidPub's Keypals: E-Mail Penpals from Around the World
http://www.kidpub.org/kidpub/keypals

Powerful African-American Images Revealed in Picture Books
http://www.scils.rutgers.edu/special/kay/afro.html

National Geographic Online
http://www.nationalgeographic.com/main.html

Virtual Frog Dissection Kit
http://george.lbl.gov/ITG.hm.pg.doc/dissect/info.html

Franklin Institute of Science Museum
http://sln.fi.edu/tfi/welcome.html

NASA (National Aeronautics and Space Administration) Home Page
http://hypatia.gsfc.nasa.gov/NASA_homepage.html

Each communications medium makes use of its own distinctive technology for gathering, encoding, sorting, and conveying its contents associated with different situations. The technological mode of a medium affects the interaction with its users, just as the method for transmitting content affects the knowledge acquired. Learning seems to be affected more by what is delivered than by the delivery system itself. In other words, the quality of the programming and the level of interactivity are the keys. All too often, computer games (and the like) encourage children to play alone. Like the television set, computer-based activities are becoming babysitters.

Processing must always take place and this requires skill. The closer the match between the way information is presented and the way it can be mentally represented, the easier it is to learn. Better communication means easier processing and more transfer. More than a decade ago, research suggested that voluntary attention and the formation of ideas can be facilitated by electronic media, with concepts becoming part of the child's repertoire (Brown, 1988). Now, new educational choices are being laid open by new electronic technologies.

Some schools are arranging for courses to be taught via the Internet. Whatever form it has taken, distance learning has always had its problems. Using satellites and television may give us the ability to distribute education more widely and meet some very specific needs, like teaching the Japanese language to three students in central Iowa, but that does not mean doing away with classrooms. The impulse to use technology to somehow increase productivity and reduce the cost of education rarely works out. Neither computers nor anything replaces good teachers. The question is how to use the technology to support these teachers and do something really worthwhile for education, culture, and society.

THE ON-LINE CLASSROOM

Internet users often make the following analogy to describe exploring the World Wide Web. The Web is a large, vaguely mapped territory with unusually beautiful scenery, unfamiliar languages and customs, and treacherous and technological jungles. For computer explorers, the Web is the most exciting communications medium since people a generation ago listened to explorers of the North Pole on their crystal radios. The Web and other equipment, such as portable computers, digital cameras, and satellite telephones, enable us to participate in rich adventures as they happen on land, sea, air, and space (Lewis, 1996).

In the early 1990s, *Mosaic* unlocked the Web for ordinary users. Then Netscape *Navigator* (the latest version is Netscape *Communicator*) came along and became one of the more popular browser tools for making the Internet even more accessible. With some websites, we use *Internet Explorer*. Such browsers are now often part of the all-in-one Internet starter kits that are taking the pain out of the process. As Web browsers are finding their way into applications programs, the necessary software for connecting to the Internet is coming already installed in many computers. The basic idea is to get away from the more cumbersome "folder and file" system. Microsoft has woven the World Wide Web into its current Windows PC operating system. Apple Computer and Netscape Communications are doing the same thing for new Macintosh computers. In both cases, the idea is to make it easier for everybody by arranging for the computer (when it comes out of the box) to plug into the phone line and handle everything as if it were a Web page.

Students can retrieve images and text from information sources arranged as World Wide Web pages by clicking the mouse on highlighted words or phrases. We are now at a stage where even novices can find their way around the global Internet network—downloading images, messages, audio, and video with the click of a mouse. The Internet has become an example of how a virtual community can connect telecommunicators around the world. Educators are increasingly looking to the cyberspace reached by these data highways as they strive to make their classrooms more interactive, collaborative, and student centered.

As we put together the technological components that provide access to a truly individualized set of active learning experiences, it is important to develop a modern philosophy of teaching, learning, and social equity. While new educational communications technology has the potential to make society more equal, it has the opposite effect if access is limited to those with the money for equipment. As we enter a world of computers, camcorders, interactive TV, satellite technology, and databases, schools are usually behind and find themselves trying to catch up with the more technologically advanced.

Electronically connecting the human mind to global information resources will result in a shift in human consciousness similar to the change that occurred when society moved from an oral to a written culture. The challenge is to make sure that this information is available for all in a 21st-century version of the public library. The technology could give us the ability to impact upon the tone and priorities of gathering information and learning in a democratic society. Of course, every technology has the potential for both freedom and domination. Like other technological advances, this one can mirror back to us all vast troves of information? Soon this technology will set up possibilities for having our best teachers available for tutoring on the Internet.

There is no doubt about the fact that the world is reengineering itself with many technological processes. The convergence of communication technologies may be one of the codes to transforming the learning process and making people more creative, resourceful, and innovative in the things they do. But do not expect technology to turn things around overnight. While learning to use what is available today, we need to start building a social and educational infrastructure that can travel the knowledge highways of the future. Experts may disagree about the ultimate consequences of innovation in electronic learning, but the development of basic skills, habits of the mind, wisdom, and traits of character will be increasingly affected by the technology.

> *One of the enduring difficulties about technology and education is that a lot of people think about the technology first and education later, if at all.*
> —Wishnietsky (1997)

NEW MEDIA REALITIES CREATED BY NETWORKING TECHNOLOGIES

Though the research is thin, there are some studies that point to some potential benefits when students and teachers use new computer-based technology and information networks. The following are some examples:

- Computer-based simulations and laboratories can be downloaded and help support national standards by involving students in active, inquiry-based learning (Dwyer, 1994).
- Networking technology, like the Internet, can help bring schools and homes closer together (Hendon, 1997).
- Technology and telecommunications can help include students with a wide range of disabilities in regular classrooms (Woronov, 1994).
- Distance learning, through networks like the Internet, can extend the learning community beyond the classroom walls (TEAMS Distance Learning, 1994).
- The Internet may help teachers continue to learn, while sharing problems and solutions with colleagues around the world (Adams & Hamm, 1996).

Since the Internet is rarely censored, it is important to supervise student work or use a program that blocks adult concerns. We suggest that teachers keep an eye on what students are doing and make sure that the classroom is off-line when a substitute teacher is in. Censorship is not the answer. There are plenty of computer programs for the Internet that act like the V-chip in new television sets. If properly arranged, the technology can protect children from the cretins and nasty folks out there.

You should not and cannot reduce everything in the world to the intellectual level of a seven-year-old. What we need is a little intellectual depth for those of all ages who choose it. A program like *Internet Nanny* is another way to prevent children from accessing inappropriate material. Just as with libraries and bookstores, it is important not to restrict the free flow of information and ideas.

To figure out how the technology works, your class can use a search engine even if you do not know how. Remember that an upper-grade class will have a few students who can get around the Internet and even build a simple website. Students really like teaching other students how to do these things. The teacher can watch and learn. Even the most experienced computer educator cannot keep up with a hundred new programs coming out every few months. The solution is to tell students to critique a new program, teach it to other students, and then teach the teacher.

In today's world, children grow up interacting with electronic media as much as they do interacting with print or people. They are engaged. But does this mean that they are learning anything meaningful or that they are making good use of either educational or leisure time? The Internet, like other electronic media, can distract students from direct interaction with peers, inhibiting important group, literacy, and physical-exercise activities. The future may be bumpy, but it does not have to be gloomy. Good use of any learning tool depends on the strength and capacity of teachers. The best results occur when it is informed educators who are driving change rather than the technology itself.

Sailing through the crosscurrents of a technological age and harmonizing the present and the future means more than reinventing the schools. It means attending to support mechanisms and the development of habits of the heart and habits of the intellect. Thinking about the educational process has to proceed thinking about the technology.

Rapid changes in information technology is resulting in less and less of a difference between the television screen, the computer screen, and telephone-linked networks. In fact, connecting a cable modem to the cable TV line (if the cable company does its part) results in much faster communication and quick acting full-motion video. When on-line is on cable it becomes even easier to swap e-mail, participate in electronic chat groups, and quickly roam around the Web. Within a few years, very broad bandwidths will result in much faster connections.

NEW TECHNOLOGIES, LEARNING COMMUNITIES, AND EDUCATIONAL PRIORITIES

The Internet is just one example of how linked multimedia computers can help us weave a new community—or waste a great deal of time. If linking the classroom to the Internet is going to have positive results, we need a clear set of educational priorities before we select the technologies to advance those priorities. In addition, teachers need training and sustained support to properly harness the technology to instructional purposes. The notion that positive things happen by simply putting the technology in the classrooms and connecting to the Internet is wrong. The technology only helps children learn better if it is part of an overall learning strategy.

It is important to remember that learning works best within the context of human relationships. When it comes to using educational technology wisely, we would do well to remember that just about everything that happens in the classroom must be filtered through the mind of the teacher. In recognition of this fact, the National Educational Goals Panel recognizes the need for "the nation's teaching force to have access to continual improvement of their professional skills" (1994). As technological potential and hazard intrude on our schools, there is general agreement that teachers need high-tech inservice training to deal with the explosion of electronic possibility.

As teachers look for ways to engage students with technology, they must ask themselves, "What is the problem to which this can be applied?" It may sometimes be a Faustian bargain; something important is given and something important taken away. The quality of the technological content, the connection to important subject matter, and a recognition of the characteristics of effective instruction are central factors in determining instructional success. School is, after all, a place where students should come into contact with caring adults and learn to work together in groups.

CONVERGING TECHNOLOGIES

The convergence of broadcast television, cable, computer, telephone, videogames, educational software, and publishing offers opportunities to entertain for profit, to inform, and to educate. The explosion of technological advances can enslave our youth with mindless videogames (in an attempt to stave off boredom) or empower them to learn and to think in new and interesting ways. Even leisure time should be held to a higher standard than "killing time." It should concern itself with replenishing the spirit: for example, a discussion of ideas, attending a fine arts event, watching good films or theater, listening to music, or hiking. According to the Centers for Disease Control and Prevention, nearly twice as many students were not active in the 1990s as compared to the mid-1980s. The same study found correlations between economic class, television watching time, weight gain, and a lack of exercise. The void left when the homework is finished and the chores are done is often filled by passively sitting in front of the television set. The result is a lot of obese and unhealthy children. Why not raise expectations and use the communications tools at hand to enhance healthy physical recreation, the enjoyment of the arts, and lifelong learning. Is it possible to structure new media systems so that they enable students to become active, intelligent, and informed citizens?

On the eve of the 21st century, there is a race in the communications industry to merge two-way telecommunications, computers, and television in order to offer new digital entertainment and information services. Satellite dishes the size of a pizza are already competitive with the cable TV industry, and new interactive TV services are being provided by regional telephone companies. The process will redefine the nation's watching and thinking habits. Fiber optics, digital transmission, and a trove of multimedia choices will change not only what Americans experience but how they experience it. A major focus should be on how to provide more choice and how to make the interactive experience just as new and exciting as when television and computers first came out.

Through the integration of computer and television technology, the concept of application will be replaced with the concept of channel. In other words, television "viewers" will become "users." More specifically, this integration will enable the user to store, control, and electronically travel through a vast array of choices. In the future, the system could include an "intelligent assistant" to record programs the user might want—without being told. Microsoft's *Bob* is an example of a virtual desktop program that simulates rooms in a house. The PC user gets the program moving by clicking on familiar objects in a room. It is simple, easy to use, and uses cartoon-like characters to help write letters, balance the checkbook, and send 200 instant e-mail (junk mail) messages. The cartoon figures can learn an individual user's level of expertise, preferences, needs, and what messages get priority. These graphical interface features guides you through a user-friendly multimedia environment that avoids complex keyboard commands. This new technology not only can make information more visually intriguing, but also can provide two-way communication with live or artificially intelligent experts. Tutoring, for example, can be done by live, recorded, or computer-composed experts.

There is a need for user-friendly electronic program guides to help customers navigate the expanding and confusing market of communications and multimedia software. Writing will continue to be important. A bad example is the typical software manuals written by a computer expert. They usually fail to clearly communicate to the uninitiated, because the writers lack training in boiling down intricate operations into concise and easily understandable steps. Articulate and highly trained writers who have taught inexperienced customers new software programs would be the best candidates for composing such manuals. To help them develop clear technological writing skills, we have had students try their hand at rewriting some of the technological directions they come across, and we occasionally send the results to the hardware or software manufacturer.

When it comes to regular classroom assignments, computers can be a great aid for helping children learn to write well across the disciplines. At the pre-writing stage, students can gather information and stimulate possibilities; during the writing phase, a sharing social context can affect their awareness of how their work communicates with others; and during the revision phase, they learn to revise and edit their early drafts. Motivation is also enhanced and the final sharing (publication) effort is bound to be more professional looking.

PROMISE, PITFALLS, AND SOCIAL EFFECTS

The two founding fathers most interested in education, Thomas Jefferson and Benjamin Franklin, both believed in spreading schooling widely among the people. Franklin wanted to train students to enter the world of work. Jefferson believed in teaching students how to work out their greatest happiness as citizens (a much wider notion than personal happiness).

The purpose of the public schools is to do the following:

- Promote the joy of learning for its own sake.
- Promote large inclusive narratives for all students to believe in.
- Be preparatory arenas for civic life.
- Help students enter the economic world with concrete workplace skills.

The tension of these sometimes conflicting purposes is still with us. But whatever combination of views you accept, our educational structures need to connect with changing media realities.

A central fact of the last twenty years is that technological change and globalization have enriched the most powerful while severely hurting the less-educated half of the American population. In the *End of Work*, Jeremy Rifkin (1996, March) writes about the danger of social unrest as we get a growing class of technological "have nots" who are more separated than ever from a core of knowledge workers. As retrenchment in social and educational spending takes place, a technological underclass is growing up that is outside of the bounds of common humanity. Worse yet, our obligations to each other are no longer reflected by our government and other social and educational institutions.

Technological powerlessness can add to cumulative social neglect and exacerbate the aftereffects of social discontent. The process should be enough to send shivers down the national spine. We are not just witnessing sins of omission and commission, but a reminder of how cumulative neglect can harm everything from character to altruism and compassion. Recent surveys have shown that learning on-line has become yet another socioeducational fault line, with minority and low-income students less likely than students in wealthier school districts to have classroom access to computers and the Internet. What is needed is a new, technologically connected vision of excellence in a vital new curriculum that includes all children (Wishnietsky, 1997).

We have to ask some basic questions about the values inherent in any new technology. What social or educational problems is it going to solve and how will it create other problems that we might prepare for. As we sort out technological myths from promising possibilities, the challenge is to arrange the technology so that it will help children manage both their technological and social lives with intelligence. How can we go about inventing a technologically intensive education that inculcates basic social competencies such as cooperation, self-control, self-awareness, reasoning, and empathy?

Technological innovations are bound to influence how we shape our social connections and how we develop socially responsible behavior. As Richard Sclove points out in *Democracy and Technology*, "People are prone to resign themselves to social circumstances established through technological artifice and practices that they might well reject if the same results were proposed through a formal political process" (1996, p. 56). When it comes to the schools, using any technology needs to be part of an overall learning strategy or it will not enlarge educational opportunities or cultivate social responsibilities.

What happens when students spend more and more time in cyberspace? Everyone enters school a little like an exposed photographic negative. Education can develop and focus the possibilities in almost any direction. Social intelligence and traditional academics are all part of the mix. Whether it is Albert Speer's Nazi Germany or Robert McNamara's Vietnam, being the best and the brightest does not guarantee high moral values or great wisdom. Social intelligence and empathy are also needed. Justice Oliver Wendell Holmes said about Franklin Roosevelt, that he had a second class intellect and a first class temperament. Developing social intelligence is central to democracy and can contribute to academics while adding to the traditional aspects of intelligence. The moral dimensions of teaching really are important. Making sure that social responsibility and civic competence is part of students' daily lives in school means expanding our vision of the curriculum and revisiting some of the classical roles assigned to education in the past.

Do technologies like the Internet encourage a sense of community or push users to splinter into discrete groups? Computer-controlled networks can help draw us together or scatter us into a million electronic communities that isolates us from others. We can gain access to badly needed information or get buried under an avalanche of information so deep that it blurs what is worth remembering. Children are even more vulnerable to what may be destructive or enriching. What does it mean to reintroduce social responsibility into the technologically intensive educational environment of the late 20th century?

Teachers can contribute to helping students develop social responsibility when they teach many subjects by incorporating social skills like cooperation, group communication, and a sense of living in a learning community. They can also work with students to create ground rules for treating others with caring and respect. The Internet may help by merging public issues with citizenship education and personal development, and helping young people recognize the importance of a life of contribution to the public good (Snauwaert, 1993). Technology can redefine our concept of the social group. Membership in an electronic group may confer the status of "citizen of the world" or make the recipient the final dumping ground for a glut of e-mail. When we get 200 e-mail messages in a day (at work), we just erase them all (without reading) and assume that if it is important someone will contact us directly.

What will it take to open a vast array of new high-tech learning possibilities for children? Promoting high technology will not help much if American schools are in a state of disrepair. Covering the playground with "temporary" trailer shacks just will not do. We need to recognize the fact that the educational structures and lessons that our machines fit into are more important than the technology itself. Even the best schools have just started shaping teaching materials for the Internet. Teachers are beginning to learn how to steer students toward well-organized websites. As many are finding out, motivation is one thing, bringing the Internet into the core of the curriculum quite another.

Today's technological and marketplace changes in communications are sweeping and profound. In many respects the Internet is becoming part of a new and larger canvas, a more personalized two-way street that is altering how we learn, work, play, and live. We just have to learn how to sort out the promise from the pitfalls and not assume that either computers or the Internet guarantee good schooling. Technological literacy is viewed by the public as important. They are right.

Elements of electronic learning most certainly will be part of our future. And if the schools do not participate in shaping new media, then they are bound to be shaped by it.

CONNECTING STUDENTS TO A CHANGING TECHNOLOGICAL WORLD

Science helps drive technology and technology returns the favor. Technology expands as science and mathematics call for more sophisticated instrumentation and techniques to study phenomena that are unobservable by other means due to danger, quantity, speed, size, or distance. As technology provides tools for investigations of the natural world, it expands scientific knowledge beyond preset boundaries. The child's natural world is not neatly divided up into disciplines, and teachers can use all the technological help that they can get to soften calcified subject-matter boundaries. The appropriate technological instruments can help you with cross-disciplinary themes while engaging students in a study of the physical and biological universe.

Technology is altering how we learn, play, live, and work. But if our faith in technology becomes a powerful ideology we miss the point. It can be magical, but if digital learning is to be healthy than we have to ask challenging questions about it. A little skepticism will improve the product. Experienced teachers know that educational shortcuts from filmstrips to videotapes have promised a lot and delivered a little. To paraphrase Jane Austen, when unquestioned vanity goes to work on a weak mind it produces every kind of mischief. This does not mean driving off into a trackless wasteland just because the information highway is full of potholes. Though rigorous thought is not its current strength, educational technology still has a lot to offer. The drab reality of spending hours at a keyboard will become more interesting. But it will still take critical thinking, commitment, social interaction, and hard work to teach and learn.

A book, *Beyond the Classroom* (Steinberg, Brown, & Dornbusch, 1996), points to broad societal problems that are holding our schools back. This study of more than 20 thousand students from diverse backgrounds found that after the elementary grades the majority were "disengaged" and just going through the motions at school. The book also implies that being high-tech is not going to get it done when fewer than one in five American students say that their friends think that it is important to get good grades. Even those coming from homes where parents stress achievement have their values undermined when the subculture scorns academic achievement. With all of the high-tech explosion of possibilities, it is important to remember that the curriculum connections to the natural world must be filtered through the mind of the student and the mind of the teacher. Investment in "human ware" beats investment in "software" every time. The professional development of teachers needs to be a top priority. School administrators must pay close attention to the barriers that prevent teachers from using available technology effectively. One good way to help is to have a help line for on-site support when computers or software will not work. Like all superhighways, the information superhighway comes with a price. It costs more money and much of the scenery is reduced to a blur.

THE FUTURE OF LEARNING IN CYBERSPACE

Harmonizing the educational present and the educational future means reinventing the schools, attending to support mechanisms, and having the courage to be out in front on certain issues. Every age seeks out the appropriate medium or combination of media to confront the mysteries of human learning. The computer is the represenional medium that can be an animated wonderland, an interactive book, a theatre, a sports arena, and even a potential life-form. Along with its associates, it is already providing for the rapid dissemination of ideas and giving us a new stage for participatory experiences.

In today's world, children grow up interacting with electronic media as much as they do interacting with print or people. Does being engaged by electronic media mean that children are making good use of leisure time or learning anything meaningful? The future may be bumpy, but it does not have to be gloomy. The technology does open up some possibilities. Parents, teachers, and programming are the foundations on which possibilities can be built. In schools, the technology can also play a role in reexamining what to teach and how to teach it. It can end the isolation of teacher and student as more and more people spend time in cyberspace. The growing power of the Internet (and other networks like America Online and corporate intranets) will make everything different. National boundaries are less noticeable and students can be heard around the world. A cornucopia of information is available for the taking. Some of it may be fraudulent, but the excitement of discovery makes it very interesting to young people.

The trouble with the future is that there are so many of them. Just because students like certain aspects of electronic learning is no guarantee that they are learning a great deal. New communication technologies can be used to help students understand imagery, solve problems creatively, and apply these solutions to real-life situations, but it takes effort and planning to make the engaging meaningful. Having a solid educational agenda is more important than having the greatest technology, but once the education piece is in place it is time to figure out exactly how technology can help you.

Schools are caught between the promise and the reality of educational technology. Simply by causing us to rethink the content of the curriculum, computers and their technological associates will contribute to educational reform. Of course, technology can do much more. It can add power to the curriculum, amplify learning, and give us more possibilities for collaborative engagement. In the long run, technological tools are rapidly becoming powerful levers in the hands of thoughtful teachers. Whatever the curriculum, it ultimately comes down to the teacher's intellectual curiosity, sense of humor, character, and ability to relate to young people. When the knowledge of effective instruction (process) is combined with subject-matter expertise (content), powerful possibilities arise.

As a close associate of instruction in tomorrow's schools, educational technology will dramatically change—and change again—how teachers go about their work. The historian Arnold Toynbee once explained the flexibility needed to advance any field as learning each day what he needed to know to do the next day's work. Science, learning theory, and technology move ahead in much the same way. After each hard-won insight, the pause should just be long enough to plot a new course designed to take advantage of what has just been learned.

> *The year's doors open*
> *like those of language,*
> *toward the unknown.*
> *Last night you told me:*
> > *tomorrow*
> *we shall have to think up signs,*
> *sketch a landscape, fabricate a plan*
> *on the double page*
> *of day and paper.*
> *Tomorrow, we shall have to invent,*
> *once more,*
> *the reality of this world.*
> > —Octavio Paz (1979)
> > translated by Elizabeth Bishop

References

Activities to Integrate Mathematics and Science (AIMS). (1987). *Glide into winter with math and science*. Fresno, CA: AIMS Education Foundation.
Adams, D. (1984). What children read influences how they write. *The Leaflet*, NCTE.
Adams, D., & Hamm, M. (1989). *Media and literacy: Learning in an electronic age*. Springfield, IL: Charles Thomas.
Adams, D., & Hamm, M. (1996). *Collaborative inquiry in science, math, and technology*. Portsmouth, NH: Heinemann.
Allende, I. (1995, February). Writing fiction. Speech given at Stanford University, Palo Alto, CA.
Allington, R. (1983, May). The reading instruction provided readers of differing reading abilities. *Elementary School Journal*.
American Association for the Advancement of Science. (AAAS) (1990). *Science for all Americans*. Washington, DC: Author.
Anderson, L., & Burns, R. (1989). *Research in the classroom: The study of teachers, teaching, and instruction*. Oxford, UK: Paramon.
Angeles, R., Juniu, S., Birnbaum, S., & Maughm, G. (1998). *Web sites for teachers*. Research project, Montclair State University, Upper Montclair, NJ.
Applebee, A., Langer, J., Mullis, I., & Jenkins, L. (1990). *The writing report card*. Princeton, NJ: Educational Testing Service.
Aquinas, T. (1952). *Summa theologica*. Trans. D. J. Sullivan. Chicago: University of Chicago Press.
Armstrong, J. (1994). *Little salt lick and the sun king*. New York: Crown.
Bang, M. (1991). *Ten, nine, eight . . .* New York: Scholastic.
Barba, R. (1995). *Science in the multicultural classroom*. Needham Heights, MA: Allyn and Bacon.
Baron, J. (1988). *Thinking and deciding*. New York: Cambridge University Press.

Barr, R., Kamil, M. L., Rosenthal, B. P., & Pearson, P. D. (Eds.). (1993). *Handbook of reading research*. White Plains, NY: Longman.

Berman, L., Hultgam, F., Lee, D., Rivkin, M., & Roderick, J. (1991). *Towards a curriculum for being*. Albany: State University of New York Press.

Bianculli, D. (1992). *Taking television seriously*. New York: Continuum.

Blackburn, K., & Lammers, J. (1996). *Kids paper airplane book*. New York: Workman.

Boomer, G. (1988). *Negotiating the curriculum*. Basingstake, UK: Falmer.

Borges, J. L. (1996). Quoted in National Council for Teacher of English and International Reading Association, *Standards for the Language Arts*. Newark, DE: National Council of Teachers of English and International Reading Association.

Botel, M., & Lytle, S. (1990). *PCRP II*. Harrisburg: Pennsylvania Department of Education.

Brown, A. L. (1988). *Metacognitive skills and reading*. White Plains, NY: Longman.

Burns, P., Roe, B., & Ross, E. (1988). *Teaching reading in today's elementary schools*. Boston: Houghton Mifflin.

Bybee, R. W., & DeBoer, G. E. (1994). Research on goals for the science curriculum. In D. Gable (Ed.), *Handbook of research in science teaching and learning* (pp. 26–28). New York: Macmillan.

Calkins, L. M. (1994). *The art of teaching writing*. Portsmouth, NH: Heinemann.

Carle, E. (n.d.). *The very hungry caterpillar*. New York: Putnam.

CCSSO (Council of Chief State School Officers). (1994). *Annual report*. Washington, DC: Author.

Centers for Disease Control and Preventon (CDC). (1998). *CDC prevention guidelines*. Atlanta: CDC. netinfo@cdc.gov

Children Now. (1996, August). *Children watching television: The role of advertisers*. Oakland, CA: Children Now. E-mail: Children Now@childrennow.org

Clemens, P. (1996). *Updated super wings: The step-by-step paper airplane book*. LosAngeles: RGA Publishing Group.

Cohen, E. G., Lotan, R., & Catanzarite, L. (1990). Treating status problems in the cooperative classroom. In S. Sharan (Ed.), *Cooperative learning: Theory and research* (pp. 203–229). New York: Praeger.

Csikszentmihalyi, M. (1996). *Creativity: Flow and the psychology of discovery and invention*. New York: HarperCollins.

Cuban, L. (1994). Neoprogressive visions and organizational realities. In *Visions for the use of computers in classroom instruction: Symposium and response*. Cambridge, MA: Harvard Educational Review Reprint, Harvard Graduate School of Education.

Daniels, H. (1994). *Literature circles: Voice and choice in the student-centered classroom*. York, ME: Stenhouse.

Dede, C. J. (1985). Assessing the potential of educational information utilities. *Library Hi Tech, 3* (4), 115–119.

Denman, G. (1989). *When you've made it your own: Teaching poetry to young people*. Portsmouth, NH: Heinemann.

Dewey, J. (1916). *Democracy and education*. New York: Macmillan.

Dissanayake, E. (1992). *Homo Aestheticus.* New York: Free Press.
Duckworth, E. (1987). *The having of wonderful ideas and other essays on teaching and learning.* New York: Teachers College Press.
Dwyer, D. (1994, April). Apple classrooms of tomorrow: What we've learned. *Educational Leadership, 51* (7), 4–10.
Eliot, T. S. (Ed.). (1975). *Epigraph "Hamlet."* New York: Harcourt Brace Jovanovich.
Evens, W. (1996). Science and reason in film and television. *Scientific Inquirer* (January–February), 45–58.
Fields, S. (1988). Coping: A cooperative learning strategy for all students. *Science Scope, 12* (3), 12–14.
Gardner, H. (1982). *Art, mind and brain: A cognitive approach to creativity.* New York: Basic Books.
Gardner, H. (1983). *Frames of mind.* New York: Basic Books.
Gardner, H. (1990). *To open minds.* New York: Basic Books.
Gardner, H. (1991). *The unschooled mind.* New York: Basic Books.
Gardner, H. (1993). *Creating minds.* New York: Basic Books.
Gardner, H. (1997). Multiple intelligence as a partner in school improvement. *Educational Leadership, 55* (1), 20–21.
George, G. C. (1972). *Julie of the wolves.* New York: Harper & Row.
Good, T., & Brophy, J. (1994). *Looking in classrooms* (6th ed.). New York: HarperCollins.
Goodlad, J. I. (1983). A study of schooling: Some findings and hypotheses. *Phi Delta Kappan, 64,* 465–470.
Graves, D. (1994). *A fresh look at writing.* Portsmouth, NH: Heinemann.
Hamm, M. (1993, Spring). Scientific literacy: Extending connections through science, mathematics and technology. *School of Education Review,* 62–67.
Hamm, M., & Adams, D. (1992). *The collaborative dimensions of learning.* Norwood, NJ: Ablex.
Hassard, J. (1990). *Science experiences: Cooperative learning and the teaching of science.* Menlo Park, CA: Addison-Wesley.
Heddens, J., & Speer, W. (1994). *Today's mathematics.* New York: Macmillan.
Heller, R. (n.d.). *Animals born alive and well.* New York: Putnam.
Heller, R. (n.d.). *Chickens aren't the only ones.* New York: Putnam.
Heller, R. (n.d.). *The reason for a flower.* New York: Putnam.
Hendon, R. A. (1997). The implications of internet usage for innovative education. Unpublished manuscript.
Hertz-Lazarowtiz, R., Sharan, S., & Steinberg, R. (1980). Classroom learning styles of elementary school children. *Journal of Educucational Psychology, 72* (99), 106.
Holmes, O. W. (1939). *Representative selections.* New York: American Book Co.
Hopkins, L. B. (Ed.). (1987). *A poetry anthology for children: Dinosaurs.* Selected by Lee Bennett Hopkins. Orlando, FL: Harcourt Brace & Co.
Horgan, J. (1996). *The end of science: Facing the limits of knowledge in the twilight of the scientific age.* New York: Helix/Addison-Wesley.
Houghton Mifflin. (1994). *The literature experience series.* Boston: Author.
Howe, A., & Jones, L. (1993). *Engaging children in science.* New York: Merrill.

Jones, R. M., & Johnson, L. C. (1990). Improving at-risk student scores. *Southwest Journal of Educational Research into Practice, 3*, 13–17.
Kidder, T. (1989). *Among schoolchildren.* New York: Houghton Mifflin.
Kinkle, J. (1998). *Inquiry project: Science, language & literacy.* Upper Montclair, NJ: Montclair State University.
Kraus, R. (n.d.). *The carrot seed.* New York: Scholastic.
Kuhn, T. (1962). *The structure of scientific revolutions.* Chicago: University of Chicago Press.
Kumin, M. (1984). *The microscope.* New York: Harper & Row.
Labbo, L. D., Hoffman, J. V., & Roser, N. L. (1995). Ways to unintentionally make writing difficult. *Language Arts, 72* (3), 164–170.
Lazar, A. M. (1993). *The construction of two college study groups.* Unpublished doctoral diss. University of Pennsylvania, Philadelphia.
Lewis, P. (1996, July 7). Adventures can find company on the internet. *New York Times*, p. 6.
Lindbergh, A. M. (1956). *The unicorn and other poems: 1935–55.* New York: Pantheon.
Maili, G., & Howe, A. (1979). Development of earth and gravity concepts among Nepali children. *Science Education, 63* (5), 685–691.
Manzo, A., & Manzo, U. (1997). *Content area literacy: Interactive teaching for active learning.* Columbus, OH: Merrill.
Martin, B. (1983). *Brown bear, brown bear, what do you see?* New York: Holt, Rinehart & Winston.
Martin, B., & Carle, E. (1991). *Polar bear, polar bear, what do you hear?* New York: Holt, Rinehart & Winston.
Martinello, M., & Cook, G. (1994). *Interdisciplinary inquiry in teaching and learning.* New York: Macmillan.
Marzano, R., Brandt, R., Hughes, C., Jones, B., Presssein, B., Rankin, S., & Suhor, C. (1988). *Dimensions of thinking: A framework for curriculum and instruction.* Alexandria, VA: Association for Supervision and Curriculum Development.
Mathews, L. (n.d.). *Bunches and bunches of bunnies.* New York: Scholastic.
McKibben, B. (1992). *The age of missing information.* New York: Random House.
McLeod, J. M. (1982). *Television and behavior: Ten years of scientific progress.* Washington, DC: U.S. Government Printing Office.
Mcqueen, L. (n.d.). *The little red hen.* New York: Putnam.
Mechling, K. R., & Oliver, D. L. (1983). *Handbook science teachers basic skills.* Washington, DC: National Science Teachers Association.
Miller, J. (1990). Survey data from Northern Illinois University's public opinion laboratory. Unpublished raw data.
Mitchell, R. (1989). *Portfolio newsletter of Arts Propel.* Cambridge: Harvard University Press.
Morriss, P., & Tchudi, S. (1996). *The new literacy.* San Francisco: Jossey-Bass.
Mumme, J. (1990). *Portfolio assessment in mathematics.* Santa Barbara: California Mathematics Project, University of California Press.
National assessment of educational progress. (1995). Washington, DC: U.S. Publications.

National Commission for Excellence in Education. (1993). *A nation at risk.* Washington, DC: National Commission for Excellence in Education, p. 22.
National Commission on Testing and Public Policy. (1990). *From gatekeeper to gateway: Transforming testing in America.* Chestnut Hill, MA: Author.
National Education Goals Panel. (1994). *The National Education Goals report 1993: Building a nation of learners.* Washington, DC: Author.
National Education Goals Panel. (1998, May 28). *The National Education Goals report 1998: Building a nation of learners.* Washington, DC: Author.
National Governors' Association. (1987). *Goals 2000.* National Education Goals Panel. Washington, DC: Author.
National Literacy Act. (1991). Washington, DC: U.S. Government Printing Office.
National Register of Publishing. (1998). *The standard directory of advertisers.* New Providence, NJ: Reed Elsevier.
National Research Council. (1995). *National science education standards.* Washington, DC: National Academy Press.
National Science Foundation. (1996). *Homeless in the universe.* Washington, DC: Author.
National Science Teachers' Association (NSTA). (1997). *NSTA pathways to the science standards,* Lawrence Lowery (Ed.). Arlington, VA: Author.
NCTE (National Council of Teachers of English & International Reading Association). (1996). *Standards for the English language arts.* Urbana, IL: Author.
Nell, V. (1989). *Lost in a book.* New Haven, CT: Yale University Press.
Nelson, E. (1981). *The silly song book.* Bulverde, TX: Sterling Publishing.
Neuman, D. (1993). *Experiencing elementary science.* Belmont, CA: Wadsworth.
Newlin, C. E. (1986). *New baby calf.* New York: Scholastic.
Noyes, A. (1922). *The torchbearers.* Edinburgh, Scotland: Blackwood.
Oliver, M. (1992). The summer day. From *House of light,* in *New and selected poems.* Boston: Beacon Press.
Paulos, J. (1989). *Innumeracy.* New York: Hill and Wang.
Paz, O. (1979). January first (E. Bishop, Trans.). In *Elizabeth Bishop: The complete poems 1927–1979.* New York: Farrar, Straus and Giroux.
Perkins, P. D., & Simmons, R. (1988). Patterns of misunderstandings: An integrative model of misconception in science, mathematics and programming. *Review of Educational Research, 58* (3), 303–326.
Piaget, J. (1973). *To understand is to invent: The future of education.* New York: Grossman.
Post, T., Humphreys, A, Ellis, A., & Buggey, J. (1997). *Interdisciplinary approaches to curriculum: Themes of teaching.* Columbus, OH: Merrill.
Pressley, M. (1987). What is a good strategy and why is it hard to teach? Paper presented at American Educational Research Association, Washington, DC.
Proust, M. (1981). *Remembrance of things past.* New York: Random House.
Raloff, J. (1996). When science and beliefs collide. *Science News, 149,* 360–361.
Ravitch, D. (1995). *National standards in American education.* Washington, DC: Brookings Institution.
Rhodes, L. K., & Shanklin, N. (1993). *Windows into literacy.* Portsmouth, NH: Heinemann.
Riel, M. (1989). The impact of computers in classrooms. *Journal of Research on Computing in Education, 22,* 180–190.

Rifkin, J. (1996, March). Speech given at Association for Supervision and Curriculum Development.
Rosenblatt, L. M. (1978). *The reader, the text, the poem: The transactional theory of the literary work.* Carbondale: Southern Illinois University Press.
Rosenblatt, L. M. (1993). The literary transaction: Evocation and response. In K. Holland, R. A. Hungerford, & S. B. Ernst (Eds.), *Journeying: Children responding to literature.* Portsmouth, NH: Heinemann.
Ruckman, I. (1986). *Night of the twisters.* New York: HarperCollins.
Rutherford, J., & Ahlgren, A. (1990). *Science for all Americans.* New York: Oxford University Press.
Sagan, C. (1994). *Pale blue dot.* New York: Random House.
Sagan, C. (1995). *The demon-haunted world.* New York: Random House.
Sanders, M. (1994). Technological problem-solving activities as a means of instruction: The TSM integration program. *School Science and Mathematics, 94* (1), 36–43.
Schiller, F. (1954). *Aesthetic education of man.* New Haven, CT: Yale University Press.
Schneps, M. (Author). (1988). *A private universe* [film]. Santa Monica, CA: Pyramid Film and Video.
Schubert, W., & Willis, G. (1991). *Understanding curricula and teaching through the arts.* Albany: State University of New York Press.
Schwartz, D. (1989). *How much Is a million?* Jefferson City, MO: Scholastic.
Sclove, R. (1996). *Democracy and technology.* New York: Guilford Press.
Sears, J., & Marshall, J. D. (1990). *Teaching about curriculum.* New York: Teachers College Press.
Segal, J. W., Chipman, S. F., & Glaser, R. (Eds.). (1985). *Thinking and learning skills.* Hillside, NJ: Teachers College Press.
Shamos, M. (1995). *The myth of scientific literacy.* New Brunswick, NJ: Rutgers University Press.
Sharan, S., & Shachar, R. (1988). *Language learning in the cooperative classroom.* New York: Springer-Verlag.
Short, K. G., & Pierce, K. M. (Eds.). (1990). *Talking about books: Creating literate communities.* Portsmouth, NH: Heinemann.
Sigler, R. S. (1985). *Children's thinking.* Englewood Cliffs, NJ: Prentice-Hall.
Slavin, R. E. (1989). *School and classroom organization.* Hillsdale, NJ: Lawrence Erlbaum.
Slavin, R. E. (1990). *Cooperative learning: Theory, research, and practice.* Englewood Cliffs, NJ: Prentice-Hall.
Smith, F. (1986). *Insult to intelligence: The bureaucratic invasion of our classrooms.* Portsmouth, NH: Heinemann.
Smith, F. (1996). *Standards for the English language arts.* Urbana, IL: National Council of Teachers of English.
Snauwaert, D. T. (1993). *Democracy, education, and governance: A developmental conception.* Albany: State University of New York Press.
Sparks, S. (1997). Speech given at Lawrence Livermore Laboratory, Berkeley, CA.

Steinberg, L., Brown, B., & Dornbusch, S. (1996). *Beyond the classroom: Why school reform has failed and what parents need to do.* New York: Simon & Schuster.

Steinbrink, R. E., & Stahl, R. J. (1994). *Cooperative learning in social studies: A handbook for teachers.* Menlo Park, CA: Addison-Wesley.

Stenmark, J. K. (1989). *Assessment alternatives in mathematics: An overview of assessment techniques that promote learning.* Berkeley: University of California, Lawrence Hall of Science.

Stewig, J. W., & Buege, C. (1994). *Dramatizing literature in whole language classrooms.* New York: Teachers College Press.

Stiggins, R. (1997). *Using assessment to motivate students.* Portland, OR: Assessment Training Institute.

Sudol, D., & Sudol, P. (1995). Yet another story: Writers' workshop revisited. *Language Arts, 72* (3), 171–178.

Taylor, D. (1993). *From the child's point of view.* Portsmouth, NH: Heinemann.

TEAMS Distance Learning. (1994). Los Angeles County Office of Education, 9300 Imperial Highway, Room 250, Downey, CA 90242.

Texley, J., & Wild, A. (Eds.). (1996). *NSTA pathways to the science standards.* Arlington, VA: National Science Teachers' Association.

Tinker, B. (1991). *Thinking about science.* Cambridge: Harvard University Press.

Tinkerton, J. (n.d.). *Pumpkin, pumpkin.* New York: Scholastic.

Tobin, K. G. (1990). Research on science laboratory activities: In pursuit of better questions and answers to improve learning. *School Science and Mathematics, 90* (5), 403–418.

Toynbee, A. (1971). *Saving the future.* London: Oxford University Press.

Trowbridge, L., & Bybee, R. (1996). *Integrating mathematics and science for intermediate and middle school teachers.* Englewood Cliffs, NJ: Meryl.

Van de Walle, J. (1997). *Elementary and middle school mathematics.* White Plains, NY: Longman.

Victor, E., & Kellough, R. (1994). *Science for the elementary school* (7th ed.). New York: Macmillan.

Vygotsky, L. S. (1962). *Thought and language.* Cambridge, MA: MIT Press.

Wade, S. (1990, March). Using think alouds to assess comprehension. *The Reading Teacher,* 442–451.

Webb, N. M. (1982). Student interaction and learning in small groups. *Review of Educational Research, 52,* 421–445.

Wei Wei Cai. (1997). *Thematic units.* Unpublished manuscript.

Wertsch, J. V. (1991). *Voices of the mind: A sociocultural approach to mediated action.* Cambridge: Harvard University Press.

Whitman, W. (1955) The noiseless patient spider. In O. Williams (Ed.). *The new pocket anthology of American verse.* New York: Washington Square Press.

Whyte, D. (1994). *The heart aroused.* New York: Currency Doubleday.

Wishnietsky, D. H. (1997). *Internet basics.* Bloomington, IN: Phi Delta Kappa Educational Foundation.

Wordsworth, W. (1967). The prelude. In D. Perkins (Ed.), *English romantic writers.* New York: Harcourt Brace Jovanovich.

Woronov, T. (1994). Assistive technology for literacy produces impressive results for the disabled. *The Harvard Education Letter, 10* (5), 6–7.

Zillman, D. (Ed.). (1992). *Media, children, and the family: Social scientific, psychodynamic, and clinical perspectives*. Hillsdale, NJ: Lawrence Erlbaum.

Zubrowski, B. (1979). *Bubbles: A children museum activity book*. Boston: Little, Brown.

Index

Academic standards, 4
Air:
   pressure, 140–141
   properties of, 139–142
   in space, 139–140
   water in, 139
   weight of, 141
Assessment, 53–65
   activity risky business, 63–64
   authentic, 62–63
   performance, 63–65
   portfolio, 54, 56–59
   portfolio exam, 60–61
   purposeful, 55
   rules for, 61, 62
   of student understanding, 53
   of thinking, 27
Attitudes:
   cooperative learning, 49

Big books, 107–108
   activities using, 108
   examples in science and math, 108–109
   in shared readings, 108–109
Bookmarks, 108–109
Bridge building, 13–15
Bubble making, 86–87
Buttons and shells, 82

Communicating about science, 76
Communication skills and pedagogy, 11
Comprehension, 110
   sketch a concept, 110–112
Computer software, 171, 172
Connecting students to a changing technological world, 186–188
Constructivism, 1–2
Cooperative inquiry activities, 40–45
   art, science, and language-arts connections, 44, 45
   seed hunt, 43, 44
   wind and water, 41–43
Cooperative learning, 33–52
   benefits, 49–52
   connecting standards to, 34, 35
   elements of, 33
   group investigation model, 38, 39
   group lesson, 39, 40
   research on, 45–48
   shared community in, 34
   strategies and reflection, 10
   and student attitudes, 49

Cooperative learning (*continued*)
  student roles in, 46
  teacher's role in, 38
Creative drama, 124–126
  activities, 125–127
Critical and creative thinking, 17–31
  and curriculum standards, 17, 18
  and the future, 30, 31
  and scientific literacy, 11, 12, 17, 18
  skills of, 17

Demonstrating the behavior of molecules, 83
Draw a scientist, 93
DRTA (directed reading–thinking activity), 104

Educational technology, 157–188
  changes in education, 162, 163
  changing technology, 163, 164
  mystery or magic, 158
  practical activities, 160, 161; communications time line, 161; comparative analysis, 161, 162; designing a model city, 161; the egg drop, 160
Electricity, 153–155
  current electricity, 154, 155
  simple electric circuit activity, 155
  static electricity, 153
Energy, 150–155
  bulletin board energy test, 155
  electrictiy, 154–155
  heat, 150
  light and color, 150
  sound, 151
Estimating, 80–81
Evaluating, 119, 120
Experimenting, 81

Force, friction, and simple machines, 148–149

Graphing, 77, 78

Inferring, 81

Information networks, 179–182
  benefits, 179
Inquiry process skills, 9, 11, 12, 75–81
  activities, 82–89; bubbles, 86, 87; buttons and shells, 82; demonstrating the behavior of molecules, 83; mystery liquids, 82; paper airplanes, 85; popcorn activity, 83, 84
  communicating, 76
  comparing, 75
  estimating, 80, 81
  experimenting, 81
  graphing, 77, 78
  inferring, 81
  measuring, 76
  observing, 75
  predicting, 79, 80
  reviewing the process skills, 88
  sequencing, 76
  sharing, 79
  sorting and classifying, 75
  space time relationships, 79
  using data, 77
Interdisciplinary activity, 118
Interdisciplinary inquiry:
  national standards, 16
  themes, 5, 6
Interdisciplinary thematic units, 130–133
  sample thematic units, 134–136; cookie mining, 136, 137; earth science, 134–143; rocks and minerals, 134–136; water, air, and weather, 137; water cohesion and surface tension, 138–143
Interdisciplinary themes, 129–156
  constancy, change, measurement, 131
  energy, 131
  evidence, models, explanations, 130
  language and language structure, 131
  mapping the future with, 155, 156
  reading, 131
  researching, 131
  systems, order, and interactions, 130

Internet technology, 172–179
  activities, 175–177
  converging technologies, 179–183
  the on-line classroom, 177–179
  web sites, 176

Language and literacy learning, 2
  collaborative approach, 97, 98
  heterogeneous groupings, 98
  write an autobiography, 127
Language and science teaching, 12, 13
  learning cycle, 15
Language arts, 95–127
  integrated, 95, 96
  literacy, 95
  literature-based approach, 99;
  standards, 5, 6
Language experience approach
  (LEA), 110
Language learning:
  newspaper scavenger hunt, 111, 112
  using media, 111
Language skills, 100–101
  reading, 100
  writing, 101
Life science, 143
  activities for primary students, 144
  activities for upper elementary students, 144, 145
  outdoor field trip, 144
  seeds, 145
Literacy, 3
  and learning, 5, 16
  scientific, 2–6
Literacy circles, 101, 102, 105
  directed reading–thinking activity (DRTA), 104

Magnets, 152, 153
Mass media, 167–171
  activities, 168, 169
  becoming an intelligent consumer, 168
  in class activities, 171
  compare print and video messages, 169, 170
  critical viewing, 168, 169
  new images from old, 171
  scrapbook for clippings, 170
  use debate for critical thought, 170, 171
  use video for instruction, 171
Measuring, 76
Meteorology, 141–143
Misconceptions, 90, 91
  overcoming, 91
  pseudoscience, 90, 91
  scientific, 92
Multimedia technology, 164–167
  comprehending media messages, 167
  TV programs, 165–167
  video applications, 165
Multiple intelligences, 29, 30
Mystery liquids activity, 82

National Assessment of Educational Progress (NAEP), 8
National Science Education Standards, 71, 72
National Science Teachers' Association (NSTA) Pathways, 73, 74
Newspaper scavenger hunt, 111, 112

Observing, 75

Paper airplanes, 85
Physical sciences, 145–155
  activities, 146–155; cold sodas warming up, 147, 148; compressing molecules, 151; heat energy, 150; light energy and color, 150; oo-bleck the mystery matter, 146; representing atoms, 147; sound and energy, 151; testing the speed of sound, 152
Piaget, 70
Poetry, 121–124
  activities, 122–124
  characteristics of, 121
  lesson, 124
  reading, 121
Popcorn activity, 83, 84

Portfolios, 54–61, 119
Predicting, 79, 80
Pressley, M., 127
Problem solving, 6, 7

Questioning, 88–93
  scientific inquiry, 88
  thoughtful, 89, 90, 93

Reader's theatre, 113, 114
Reading:
  experiences after, 106, 107
  experiences before, 102, 103
  experiences during, 103, 104
  language-arts standards, 96
Reciprocal teaching, 110
Reflection, 10
Rutherford, F. J., 94

Science curriculum, 69
  creating a quality program, 92
  history, 69, 92
  and second-language learners, 72, 73
Science process skills, 75–81
Science teaching:
  engaging students in problem settings, 92
  in the twenty-first century, 94
Scientific literacy, 67–69
  making decisions, 5, 6, 11, 12
  and thinking, 12, 13
Sequencing, 76
Sharing, 79

Sorting and classifying, 75
Sound energy, 151
  activities, 151, 152
Space–time relationships, 79
SQ3R (survey, question, read, recite, review), 109
Standardized tests, 55, 56
Standards:
  language arts, 96
  and literacy, 2, 4
  and science, 70–72
  and technology, 159, 160
Story dramatization, 126

Thematic approach, 6
Thinking:
  creative, 24, 25
  critical, 25, 26; and teachers, 28, 29
  symbolic modes, 19, 20
  teaching skills, 22, 23
  think aloud questions, 24

Using data, 77

Water, 137–139
  cohesion and surface tension, 138, 139
Wei Wei Cai, 132, 133
Writing, 115–121
  cycle, 115
  peer collaboration, 117, 118
  process, 115–121; editing, 115, 117; pre-writing, 116; revision, 117, 118

ABOUT THE AUTHORS

MARY HAMM has written articles and books on science and mathematics education. She worked on the science standards and is now a professor at San Francisco State University.

DENNIS ADAMS, a former Fulbright Scholar, currently teaches at Montclair State University in New Jersey. He has taught at the elementary school level and has authored or coauthored more than a dozen books and over one hundred articles.

www.ingramcontent.com/pod-product-compliance
Lightning Source LLC
Chambersburg PA
CBHW070327230426
43663CB00011B/2237